TS
157.4   The just-in-time self test
.F58
1995

TS
157.4
.F58
1995

INFORMATION RESOURCE CENTER
MERCURY MARINE
W6250 PIONEER RD
FOND DU LAC WI 54936-1939

THE JUST-IN-TIME SELF TEST
Success through Assessment and Implementation

THE JUST-IN-TIME SELF TEST
Success through Assessment and Implementation

Dennis Fisher

IRWIN
Professional Publishing

Chicago • Bogatá • Boston • Buenos Aires • Caracas
London • Madrid • Mexico City • Sydney • Toronto

© RICHARD D. IRWIN, INC., 1995

All rights reserved. No part of this publication may be reproduced, stored in a retrieval system, or transmitted, in any form or by any means, electronic, mechanical, photocopying, recording, or otherwise, without the prior written permission of the publisher.

Senior sponsoring editor: Cynthia A. Zigmund
Project editor: Paula M. Buschman
Production supervisor: Lara Feinberg
Designer: Heidi J. Baughman
Manager, graphics and
 desktop services: Kim Meriwether
Art and book production: Boston Graphics, Inc.
Typeface: 11/13 Palatino
Printer: Quebecor Printing Book Group/Kingsport

Library of Congress Cataloging-in-Publication Data

Fisher, Dennis
 The just-in-time self test : success through assessment and implementation : a guide for evaluating and developing a just-in -time implementation plan / Dennis Fisher.
  p.   cm.
 Includes bibliographical references and index.
 ISBN 0-7863-0299-2
  1. Just-in-time systems—Planning. I. Title.
 TS157.4.F58. 1995
 658.5'6—dc20   94-38583

Printed in the United States of America
1 2 3 4 5 6 7 8 9 0 Q/K 2 1 0 9 8 7 6 5

Dedication

This book is dedicated to my lady, a loving and supportive eternal partner.

Preface

Leading practitioners and industry experts have published many recommendations for the implementation of Just-In-Time systems. Most of these publications tend to deal with abstract higher-level issues of implementing and maintaining Just-In-Time (JIT) systems. However, these publications don't contain a single top to bottom how-to plan or an objective way of evaluating yourself. This is the purpose of *The Just-In-Time Self Test*. It is the result of many requests from manufacturers and consultants preparing for a JIT implementation who have been frustrated by the many bits and pieces of JIT data scattered throughout many books, proceedings, journals, and trade publications. *The Just-In-Time Self Test* complements but does not replace these other publications. It has three central purposes: (1) to provide a simple method of testing Just-In-Time performance; (2) to provide step-by-step guidelines to help any company to identify their operational shortcomings: and (3) to provide a primer of overall concepts, techniques, and formats that condense the majority of the hints from most of the current JIT publications into a single volume.

Another purpose for the publication of this book is to help small-to medium-size companies that have limited funds available. Many of these companies turn to JIT manufacturing as a survival strategy. This book may inhibit hiring more expensive managers or consultants to help them. *The Just-In-Time Self Test* can be used and understood by anyone. It is not necessary to be an experienced JIT person to grasp any of the concepts in this book.

One of the basic philosophies of JIT systems is to try out new techniques and then look for the benefits. Results are often dramatic and act to drive the users to implement more techniques of JIT until the entire company is involved. This bottom-up approach has happened more often than the top-down approach because of the visibly, hands-on nature of the shop floor improvements. *The Just-In-Time Self Test* supports both approaches. It defines "how tos" for both the supervisor and worker level. Then it provides the higher-level manager with an overall planning process that integrates the decision-making process to the physical product flows. Ultimately JIT will be more successful when it is implemented from the top down, but a bottom-up start can be used to show real performance to upper-level management and convince the managers that it really works and that they should support the plan. A self-evaluation based on an objective test like the JIT Self Test makes a great argument.

Finally, and most important, my special and heartfelt thanks to my wife Sandy for giving up the time she needed from me while I was writing this book. My thanks also to my professional friends and associates who contributed their experience and expertise to this work: Richard Chamberlain, Alan Dunn, Cheryl La Vella, Jack Conrad, Paul Funk, Gus Berger (posthumously), Sue Neff, John Parr, Keith Laughton, the crews at Safetran Systems and at McDonnell Douglas Computer Systems, and the many other kind people who helped me do the research.

Dennis Fisher

Contents

INTRODUCTION ... 1

Chapter One ... 2
HOW TO USE THIS BOOK

Chapter Two ... 6
IDENTIFYING YOUR JUST-IN-TIME ENVIRONMENT

TAKING THE TEST ... 11

Chapter Three ... 12
THE JUST-IN-TIME SELF TEST

Chapter Four ... 18
SCORING THE JUST-IN-TIME SELF TEST

USING THE SELF TEST FOR IMPROVEMENT ... 23

Chapter Five ... 24
EDUCATION AND PEOPLE

Chapter Six ... 38
QUALITY MANAGEMENT

Chapter Seven ... 71
FACTORY FLOW

Chapter Eight ... 82
PRODUCTION PROCESSES

Chapter Nine ... 100
MASTER PLANNING

Chapter Ten 114
PURCHASING

Chapter Eleven 124
DATA INTEGRITY

Chapter Twelve 135
RESULTS

IMPLEMENTATION 149

Chapter Thirteen 150
BUILDING AN IMPLEMENTATION PLAN

GLOSSARY 165

BIBLIOGRAPHY 203

ADDENDUM 207

INDEX 237

INTRODUCTION

Chapter One

How to Use This Book

Let's start with the definition for Just-In-Time manufacturing—a philosophy of manufacturing based on planned elimination of all wasted resources and continuous improvement of productivity. The primary rules are to have *only* the required minimum amount of top quality inventory in the proper place at the exact time when needed. This involves all activities in all departments necessary to maintain the flow of materials through the company. Other terms might be stockless production, zero inventories, or short cycle manufacturing.

The flow and design of this book will help you plot a course for immediate action. All you have to do is follow steps 1 through 6. The purpose of *The Just-In-Time Self Test* is to fulfill your need to improve by walking you through a simple step-by-step program. The following steps are the plan of this book.

1. IDENTIFY YOUR TYPE OF MANUFACTURING ENVIRONMENT

This chapter will introduce you to the different environments of repetitive manufacturing and help you identify the elements of your company that may differ from other types of Just-In-Time operations. You will be given a list of typical differences between comparative manufacturing models. Then you will:

1. Identify those elements most closely related to your operations.
2. Identify the type-model company you are most like.
3. Select the proper test from the JIT Self Tests in the addendum at the back of the book.

2. TAKE THE JUST-IN-TIME SELF TEST

The questions relate to the operating results or actions you are currently doing within your company. Either take the test yourself as an individual or give it to the whole company. You or your company will be given a set of questions to which you will answer with a simple "yes" or "no." Be as objective as possible, viewing your operations, people, policies, processes, and business practices from a distance. Avoid ego, pride, or bias in your answers.

3. SCORE THE TEST

Two scoring tools will help you to determine ABDC class ratings for yourselves and your company and then to compare them to the expert-based Just-In-Time performance-model in Chapter Four. This model is based on research from many companies, industry experts, and previous tests results. There is a test and associated score and rating sheet for several type-models in the Addendum. At the end of this step, you will have accomplished one or both of the following activities:

One: You will have *scored your individual test* by totaling your responses to the JIT Self Test and compared the results to the JIT expert-based model.

Two: you will have *scored your company* by giving the JIT Self Test to your company as a whole and scored the results using the JIT Self Test scoring and rating sheets.

Once you finish your scoring and rating sheets, you will discover whether you are a Class A, B, C, or D JIT organization. Everyone should know what the rating was if they have taken the test. Make sure you let them know about the narratives in Chapter Four tied to each class.

There are narratives in Chapter Four for each class that describe the typical condition and results of companies that fall into each class. When you have read the narratives for each of the rating classes, you will know from an overall point of view how well you compare to other companies that are implementing or planning to implement JIT.

4. EVALUATE THE RESULTS

The evaluation of the results will help you focus on the shortcomings that were highlighted in the Just-In-Time Self Test and stimulate action for improvement. You will identify your shortcomings by listing the "no" responses from the Just-In-Time Self Test and selecting which of the subsequent chapters to read to learn how to overcome the shortcomings.

5. READ THE CHAPTERS RELATING TO THE SHORTCOMINGS

As you read Chapters Five through Thirteen you will find flow charts, strategies, hints, rules, and tools that will help you discern what actions to take to improve your operations and overcome the shortcomings you defined in Chapter Five. At the end of each chapter is a Just-In-Time Action Plan chart with blank lines where you can make notes as you go about things you've learned from the chapter. Your notes can be used later to define a Just-In-Time Implementation Plan.

6. BUILD YOUR JUST-IN-TIME IMPLEMENTATION PLAN

Chapter Thirteen will become the basis of an action plan necessary to reach your JIT objectives. You can refer back to the notes you made at the end of each chapter and use them to define your preliminary JIT Implementation Plan. This concluding chapter will provide a technique and format for you to quickly and simply build your plan. A sample plan and various forms are provided to facilitate this task. If you follow these simple steps, you will have reviewed or learned these project management concepts and applications:

   a. How to build a project plan.
   b. How to organize and manage a JIT project team.
   c. The keys to success in meeting management and in team building.

   d. A start-up strategy checklist for a JIT implementation.
   e. Techniques for maintaining momentum and overcoming resistance.

There is an Addendum, Glossary, Index, and Bibliography at the back of the book to provide worksheets, definitions, and materials for your further learning. You and your Just-in-Time team will not get very far without further study. This book provides the key elements and measures for a JIT system, but there is much greater detail needed to perfect the many powerful JIT techniques that are referenced in this book.

This brings you to the end of the book. But the end of the book is only the beginning of the journey. Next you must turn your plans into action. After that, new actions, refinements, and continuous improvement will become the focus. Just-In-Time is merely a journey of many steps with many small accomplishments along the way. You must focus on these accomplishments from small improvements—*not* from an anticipated completion of a JIT implementation. JIT implementations never end. So emphasize and celebrate ongoing incremental milestones as the end in itself.

Chapter Two

Identifying Your Just-In-Time Environment

Some of you readers will have different environments of manufacturing. You will need help to identify the elements of your environment that may differ from other types of Just-In-Time operations. This chapter gives you a list of typical differences and comparative manufacturing models. You can then:

1. Check off those elements most closely related to you.
2. Compare the number of differing elements to a set of standard manufacturing models.
3. Determine your environment from the type-models provided.
4. Select which test to take from the Addendum.

The following is a list of various types of manufacturing type-models. Select one that fits your company. Those who most fit the repetitive list should consider Just-In-Time as soon as possible or they will lose competitive advantage.

REPETITIVE

Those who fit the repetitive list are most directly related to all of the elements of the full self test. Essentially all of the activities in a repetitive environment can use JIT.

- Make to stock or assemble to order products.
- Controllable work patterns.
- Medium to high volume.

Chapter 2/Identifying Your Just-In-Time Environment   7

- Repetitive patterns.
- Defects less than 2 percent.
- Assembly operations.
- Easy to understand processes.
- Predictable patterns.
- Planned by rates.
- Driven by daily schedules, not by work orders.
- Fixed routings.
- Process costing.
- Dispersed machinery formed into lines.
- *Examples:* food processing, electronics products, autos, con sumer products, paint, computers, and the like.

JOB SHOP

Those who fit the job shop list may still be able to use the JIT techniques well but are limited in some applications. It is the repetitive activities that cause most waste and that continuous improvement can improve most easily. Much of the activities in a job shop environment may be repetitive and by definition can use JIT.

- All products are unique made-to-order items.
- Plans and controls are by orders or lots.
- Use of work orders is unavoidable.
- Variable routings exist between products.
- Job costing is done by work order.
- Multipurpose machinery is grouped together.
- Low volumes are usual.
- Examples: R&D shops, sheet metal, molding, airplanes, custom woodworking, printing, and so on.

PROCESS FLOW

Production flow in a continuous process. Process industries are already in much of the JIT flow because there is little inventory between workstations. But other waste reduction issues will apply to their support activities.

- All products are made to stock or distribution.
- Planning done by shift or period rates.
- Fixed routings for all products.
- Fixed, dedicated equipment.
- Distinct line orientation for most products.
- High volumes are usual.
- *Examples:* pharmaceuticals, oil refining, chemicals, steel, cement, textiles, carpets, beer, and so forth.

PROJECT PRODUCTION

Project production usually involves special custom building of large units or the grouping of things like ships or housing projects.

- Make to order or assemble to order products.
- Planned by project or major assembly.
- Usually low volume.
- Semivariable routers.
- Usually a large stationary project.
- Uses subcontracting.
- *Examples:* shipbuilding, construction, specialty chemicals, supercollider–superconductor, etcetera.

There are many similarities and commonalities between all of the type-models listed above. Any of the questions in the Just-In-Time Self Test that are tied to the following list of similarities will apply to all types-models.

Similarities or common issues for all type-models are:

- People involvement.
- Customer satisfaction.
- Workplace organization.
- Training and cross-training.
- Supplier management.

- Total quality management.
- Management support.
- Problem solving.
- Preventive or productive maintenance.
- Set-up time reduction.
- Cycle time reduction.
- Signaling systems.
- Inventory reduction.
- Throughput improvement.
- Operating expense reduction.
- People involvement and team building.
- Storage and floor space reduction.
- Measurement systems.

Some unique differences, however, must be addressed for each type-model. The following list identifies questions in the Just-In-Time Self Test that may not apply to the specific type-model you might be working with. Be aware that most companies usually have elements of more than one type-model within their operations. If this is the case for you, don't exclude the question from your list. The following list is only intended to make the test less confusing for different type-models. These questions have been removed from the self test, and a separate test is provided in the Addendum for each type-model.

Please note that the following list is for convenience only. After you have taken the test you may want to see which questions were eliminated from your test. Then, you can use this list:

Job shop

Question 19	Question 34
Question 22	Question 39
Question 24	Question 51
Question 25	Question 58
Question 29	

Flow shop

Question 26

Project

Question 18	Question 24
Question 19	Question 25
Question 20	Question 39
Question 21	Question 51
Question 23	Question 58

After completing the self test go to the Addendum at the end of the book and score yourself on the score sheet and rating sheet that are designed for your type-model. One other issue must be addressed before continuing—company size. Many people taking the Just-In-Time Self Test will be from small companies. Some of the materials in the book make reference to resources, staff, and team sizes that would too large in scope for a small company. There will be a short paragraph near each occurrence like this in the book to address small company needs.

TAKING THE TEST

Chapter Three

The Just-In-Time Self Test

The purpose of this chapter is to provide you with a list of questions about the operating results or actions within your companies. You will be given a set of questions, which you will answer with a simple "yes" or "no." *If there is substantial activity or effort going on for an item, then answer "yes." If no real effort or practice exists, answer "no."* This is important to remember because some of the questions may not be able to be answered with a distinct "yes" or "no." If you are not aware of any activity by your company on a particular question, answer "no." You must answer every question.

You must act as objective observers: viewing your operations, people, policies, processes, and business practices from a distance. Be careful to avoid ego, pride, or bias in your answers—particularly if you are tied in some way to the activities being questioned. Answer the questions by placing a check mark or "x" in the yes or no column provided.

Be sure you have selected the correct test from the Addendum (at the end of book) based on your type of company as defined in Chapter Two. The list of questions here in this chapter is the master list of all questions that apply to a typical JIT repetitive type-model. The questionnaires in the Addendum have had some of the questions removed to match to their business environment. The following questions are the master list of all questions on the Just-In-Time Self Test.

THE JUST-IN-TIME SELF-TEST

Please respond as objectively as possible to each of the following elements of Just-In-Time.

Education and People

1. Management supports and participates in employee involvement programs that demonstrate trust, delegate authority, and allow autonomous decision making.
2. A continuous and formal training program is in place that includes new employee indoctrination, skills development, cross training, manufacturing principles, and JIT concepts for all employees.
3. Middle management and supervisors have been reduced and reorganized to support JIT.
4. Key managers, support personnel, and operators have been trained in Just-In-Time practices.
5. Cross-training programs have been implemented with skills tracking and evaluations.
6. Compensation and rewards for employees are based on both employee flexibility and team contribution.
7. Department or team problem-solving groups meet regularly to address and solve quality and flow problems throughout all departments.
8. All associated functional departments (e.g., engineering, purchasing, marketing and accounting) are part of problem-resolution teams.
9. Management responds to employee ideas and feeds back their responses immediately.

Quality Management

10. Management exhibits consistent support for quality procedures.
11. Inspection sequences have mostly been eliminated and quality is part of the individual operator's responsibility.
12. Quality control departments have been replaced with process audit and operator training functions.

13. Early warning statistical quality control tools are in place and are used to monitor and control quality at critical points in the process.

14. Quality errors are repaired or prevented at the source where they occur.

15. Fail-safe (poka-yoke) devices are installed at most locations where recurring human quality errors typically occur.

16. Work areas are consistently clean, organized, and free of unnecessary materials and equipment.

17. The majority of incoming materials are certified or source inspected at the suppliers.

18. Supplier quality certification and performance rating programs are in place and are continuously monitored.

Factory Flow

19. Transportation networks consistently deliver mixed loads from local and long distance sources.

20. Standardized containers with exact quantities are used between supplier and plant for over 50 percent of the volume parts.

21. Weekly or daily delivery of 80 percent or more of production materials are made to the plant and directly to the production line points of use.

22. Supplier deliveries are scheduled by the production processes demand for parts.

23. Standard containers hold exact and consistent quantities and are maintained throughout the shop.

24. Production build-and-pull quantities are calculated to support only average daily demand.

25. Shop schedules, dispatching, work orders, and expediting have been eliminated and priorities are defined by shift schedules and quantities.

26. Parts are only produced as required by demand and are built in quantities approaching one.

27. Labor is not "kept busy" by building product when not needed at the next operation.

Production Processes

28. Production lines are grouped into product family (group technology) cells or lines.
29. Processes support flexible mixed model runs with minimum material handling.
30. Manufacturing is actively involved in product and process design improvement for quality and producability.
31. Production parts have been designed to facilitate fast changeover.
32. Tooling and fixtures are available when setups and jobs begin.
33. Rapid setups are established (less than 10 minutes) for most machines and lines.
34. Cycle times of each workstation, cell, or line are matched to upstream and downstream times.
35. Manufacturing processes have been reoriented to eliminate material handling.
36. Process problems are identified and visibly signaled immediately on discovery.
37. Manufacturing engineering is located in the production area and is immediately available for problem resolution.
38. Scheduled preventive maintenance is considered an important part of production performance.

Master Planning

39. Daily rate and level schedules are used and meet due dates.
40. MRP is used for demand planning, customer committing, supplier schedules, and supplier capacity planning.
41. Management participates in the planning and replanning process and commits to a realistic capacity level.
42. Marketing promotes the demonstrated benefits of Just-In-Time to customers.
43. Production rates exactly equal demand rates, or production quantities equal demand quantities.
44. Customer on time delivery rate is 98-plus percent as committed.

Purchasing

45. Key volume suppliers are local.
46. Single-source suppliers make up greater than 50 percent of all suppliers.
47. The number of active suppliers has been reduced substantially—by 50 percent or more.
48. Buyers and suppliers are rated by supplier quality, delivery, and ongoing improvement.
49. Frequent multidepartment contacts are made between suppliers and your plant.
50. Most paperwork, material handling, transportation, and quality waste has been eliminated between supplier and plant.
51. Delivery lead time for most parts range from one day to one week.

Data Integrity

52. Inventory record accuracy is 98-plus percent or better for both stockrooms and point of use storage.
53. Bills of material accuracy 99-plus percent or better for costing needs and post deduct inventory.
54. Bills of materials flattened—structured with two or less levels in production.
55. Routing detail, methods, or assembly instructions are accurately defined and maintained by timely process flow changes.
56. Clearly defined engineering standards for costing.
57. Shipment forecast variation is plus or minus 10 percent or less by product family in the current time period.
58. Accounting systems and controls have been redesigned to work in a JIT environment.

Results

59. Operational measurements with both targets and tolerances are in place and reviewed daily by management and employees.
60. Overall manufacturing process cycle time and throughput lead times are reduced.
61. Production floor space has been substantially reduced.

62. Work-in-process inventories are continuously reducing.

63. Substantial increases in productivity or shipments per employee.

64. Substantially reduced operating expense.

65. Overall cost of quality is continually reducing.

66. Stockroom inventories are continuously reducing.

67. Substantially increased inventory turns.

68. On-time customer delivery is continuously increasing for both line-fill and order-fill rate.

Chapter Four

Scoring the Just-In-Time Self Test

Now that you've taken the test or have administered it to your company, two scoring tools will help you determine your Just-In-Time ratings. You can then compare yourself to the JIT rating scheme listed in the Addendum at the end of the book. This rating scheme is based on research from many companies, extensive research, and industry experts. Turn to the Addendum and follow the instructions on each type of "self test score and rating sheets."

Four sets of tests and their associated scoring and rating sheets in the Addendum are designed for each type-model.

1. *For an individual evaluation*

   a. Take the Just-In-Time Self Test.
   b. Total your "yes" responses to the Just-In-Time Self Test.
   c. Compare your personal results to the JIT model.
   d. Read the narratives in this chapter in your rating class.

2. *For a company evaluation*

   a. Give the Just-In-Time Self Test to your company as a group.
   b. Score the results using the self test scoring and rating sheets.
   c. Read the narratives in this chapter for your rating class.

Once the scoring and rating are done, it's time to check how you did in relation to others involved with Just-In-Time.

RATING NARRATIVES—HOW DO YOU COMPARE?

Class A

- Continuous improvement is a way of life.
- All employees are involved and supportive of minute-by-minute throughput objectives.
- Management and all subordinate groups communicate openly and freely.
- Profits are higher than the industry standards.
- Customers are fully satisfied with quality, delivery, and price.
- Manpower is cross-trained and highly flexible.
- Morale is high; commitment and motivation are strong.
- Inventory is minimal—turns from 20 to 200.
- Productivity continues to increase.
- Turnover is minimal.

Class B

- Continuous improvement is working.
- Most employees are involved and supportive of daily throughput objectives.
- Management and most support groups communicate openly and freely.
- Profits are close to the industry standards.
- Customers are generally satisfied with quality, delivery, and price.
- Manpower is being trained and more flexible.
- Morale is good; commitment and motivation are building.
- Inventory has been reduced dramatically—turns 10 to 20.
- Productivity is increasing.
- Turnover is average.

Class C

- Improvement is difficult and has considerable resistance.

- Few champions are involved and throughput objectives, if any, are not met very well.
- Management and most support groups have some communication problems.
- Profits are lower than the industry.
- Customers still have problems with quality, delivery and price.
- Manpower is not trained and is not flexible.
- Morale, commitment, and motivation are low.
- Inventory is coming down—turns below 10.
- Productivity may begin to increase.
- Turnover is higher.

Class D

- Little real improvement.
- Employees are not involved or supportive of throughput objectives.
- Management makes most decisions about process—workers not involved.
- Profits are lower than the industry.
- Customers are not satisfied with quality, delivery and price.
- Manpower is not trained and is unreliable.
- Morale is low.
- Inventory is high, turns are 4 or below.
- Productivity increases are nonexistent.
- Turnover is high.

The Class A through D narratives are based on industry averages and results taken from Just-In-Time Self Test users. They have been selected as indicators that would be affected by improvements and benefits generated by JIT activities.

Now, it's time to go through *The Just-In-Time Self Test* reference chapters and find all those items that were answered "no." Highlight each one of them on the test and begin looking up the descriptions and guidelines in the balance of Chapters Five through Thirteen in the book. The text is intended to give you some experienced guidelines for improving your JIT plans. Read each question and any associated references and make notes in the action items list at the end of each chapter.

Suggestion: Sort the "no" list by priority, based on the amount of effort and time to implement. Select a few of the easier ones to do first to gain some visible success. This will show others some benefits quickly. These actions will help you to develop your JIT action plan in Chapter Thirteen.

USING THE SELF TEST
FOR IMPROVEMENT

Chapter Five

Education and People

People make anything work. We wouldn't be here without people like mothers, friends, leaders, or coworkers. At times, mothers seem to know what's right more than the corporate leaders do. Mothers stress education and getting along within the family. Yet, our leaders in the workplace cut the training and the people whenever there is a slowdown in the business or when troubles exist. Why don't they listen to their mothers? The real power is in the people and their knowledge.

People are the only true cause of overall success of a Just-In-Time implementation. They must be trained. A training evaluation system starts with a "needs analysis" to determine what types of training your company needs.
People involvement and empowerment strategies will also be presented. At the completion of this chapter, you will have reviewed or learned the following concepts and applications.

1. The importance of education to JIT systems.
2. How to conduct a needs analysis.
3. How to set up and implement a cost-effective training program.
4. The power of cross-training.
5. The necessity of employee involvement programs.
6. Defining inter- and intra-departmental objectives for JIT.
7. Problem resolution in a JIT environment.

Remember to keep a list of the detailed items in this chapter that you find to be missing in your own operations and add them to the JIT action plan.

Question 1. Management supports and participates in employee involvement programs that demonstrate trust, delegate authority, and allow autonomous decision making.

The National Quality Award (Malcolm Baldrige Award) states that human resources management in a company worthy of national recognition must have an effective program for developing " . . . the full potential of the workforce, including management, and to maintain an environment conducive to full participation, quality leadership, and personal and organizational growth." Employee involvement and employee empowerment is at the heart of any program for developing formal long-lasting growth and continuous improvement required by JIT.

Employee involvement is essentially teamwork—for everybody. A strong teamwork-oriented company is flexible and capable of quickly forming into self-managed teams without the normal hierarchy directing them. Teamwork should be a way of life.

Teamwork starts with support from *all* management. It may seem that some tasks can be done better and faster by one person. At times this is true. But many problems cannot be resolved by one person, and the solutions may not fit everyone's needs. Teams can be used to solve company problems more effectively than individuals. Team activities will be further developed in questions 7, 8, and 9.

Employee empowerment goes hand in hand with employee involvement. Simply stated, teams must have the power to make decisions in their areas about their processes without higher-level interference. Decision power must be pushed down to the lowest levels—to the point of action. The workers must be a part of the decision-making process if they are to have a sense of ownership. When they make errors, they must be forgiven, particularly in the beginning of a JIT project or else this will inhibit innovation, creativity and risk-taking. Although it may take longer to reach a group consensus, when a good team decision is made implementation goes very quickly because everyone has bought into it ahead of time.

Question 2. A continuous and formal training program is in place, which includes new employee indoctrination, skills development, cross training, manufacturing principles, and JIT concepts for all employees.

All employees must understand their role in a JIT/total quality system and how their roles will change as quality improves. They need to know where their work fits into the big picture and how they will be interfacing with other workers. As the rules change about how material flows, they will need to be trained in all the aspects of the systems that influence them.

Formal technical training must be designed to provide the skills and function-specific knowledge to employees that will increase their job performance. One of the major tasks to be done to meet this goal is to conduct an education or training needs analysis. A needs analysis is not merely asking the employees what training they need, but it is a process that takes an initial look at jobs and functions and then determines what skills and knowledge are needed to do those jobs correctly. Group or individual interviews are conducted to identify what specific skills are needed *to do the work* and then what training or techniques could be integrated into the work to make it go better. There are five steps in a well-developed technical training program. They are defined as follows.

Step One: Analysis

a. Needs analysis identify the differences between training and nontraining problems and propose workable solutions for them. The needs are defined by identifying what the worker needs to perform or improve the work tasks. This phase includes what the training will address, what solutions it will provide, who will get trained, how long it will take, what type of programs will fit, and who will manage and present the training.

b. Training analysis determines the content of the training program. Subject matter, documentation, computer programs, and so on are gathered, organized, and planned to support the implementation phase.

Step Two: Design

a. Create an educational strategy based on the needs and training analysis; the design focuses on the solutions needed to solve company problems. It is the educational blueprint into which the educational content will be inserted.
b. Develop the format that defines the actual programs, plans, applicable learning principles, course outlines, training delivery systems, and tools.

Step Three: Development Phase

a. Materials are collected that identify the specific content, learning objectives, tests, examples, workshop instructions, instructor guides, student guides, all printed materials, and instructional needs.
b. Content is constructed and organized into the proper formats, which include all the printing, collating, and production of all physical aspects of the training programs.

Step Four: Implementation

a. Finalized schedules are published and distributed to all participants selected for training; room reservations are made; and administrative details are completed.
b. Conduct the training. Training is presented to learners.

Step Five: Evaluation

a. Survey learner satisfaction by gathering, collating, evaluating and reporting objectively given learner responses about the training.
b. Assess the learning to see if learning objectives were met. Focus on tying back to the needs analysis and to the original purpose of the training.
c. Measure performance by testing in class and on site for behavioral changes and newly exhibited job skills.

One of the often overlooked elements of a needs analysis is a new employee indoctrination training cycle. When new employees are hired they can become a constraint in the flow of materials if they are not fully indoctrinated in the concepts of total quality and Just-In-Time principles and practice. The following program can be used to take each new employee through specific activities prior to being placed within the JIT flow.

An Indoctrination Agenda. A time schedule and interview cycle can be set up that carries the new employee through all of the key persons and functions in a one to three day schedule. Higher level managers take longer and cover more people and functions, based on their new position. Some companies (very few) have the new managers actually work in several of the key positions with which they will be interfacing once they are in position.

The Interview Cycle. As the new employee moves from person to person in the schedule he or she should ask two questions at each stop: (1) "What is your job and how does it effect the flow of materials in the company?" (2) " What are your expectations of my job or position?"

The Daily Review. At the end of each day, the new employee will meet with his or her supervisor and share the notes taken during the interviews. This can be a very revealing process when outsiders share their observations.

The Result. The affect of this indoctrination is at least a 30 to 50 percent reduction in the learning curve cycle. Usually, a new employee takes five months to reach a clear picture of how the whole organization goes together. The indoctrination cycle can reduce the time to three months or less. Certain other benefits can arise out of the daily review process.

Small companies may approach these techniques on a much smaller scale. Most of the training might be defined by the key members and conducted as a group on an off-hour schedule. Hiring a professional trainer to conduct the training is expedient as long as someone inside has been well trained and retains the materials.

> Question 3. Middle management and supervisors have been reduced and reorganized to support JIT.

When responsibility for decision making is delegated downward to individuals and teams, the need for middle managers and supervisors is diminished. This prospect may not be pleasant, so there is usually resistance from middle management and supervisors because they feel their jobs will be threatened—and rightly so. They have an opportunity to facilitate the changes or be seen as resisters. The facilitators should be kept, and the resisters transferred or released. That is a tough decision to make. Usually, the resisters get very uncomfortable and choose to transfer out of the system. If not, they tend to get pushed aside. Essentially, the supervisors and managers must become facilitators, analysts, coordinators, change agents, and people-oriented leaders. Obviously, the resisters should be helped to join the JIT program.

> Question 4. Key managers, support personnel, and operators have been trained in Just-In-Time practices.

Just-In-Time training is fairly easy to do. The best way is to use mock products and processes from the company itself. Build a JIT demonstration and introduce it to all employees in the company, with a set of the basic JIT concepts as an introduction. It has been discovered that the most difficult groups to fully grasp the mechanical aspects of JIT are the middle to top management groups. Even the results of the Just-In-Time Self Test show the middle management groups as the most negative.

Recent interviews with a West Coast aerospace manufacturer disclosed that the ability to communicate on specific issues was visibly enhanced through training courses. The most noticeable effect was that those who had resisted cooperative interaction were seen to be supporters within only a few months.

There is more information on project management and organization in Chapter Thirteen—Implementation.

> Question 5. Cross-training programs have been implemented with skills tracking and evaluations.

People are portable capacity. They represent considerable power when rapid change occurs. A good program of cross training adds power and flexibility to a JIT implementation.

Guidelines for the National Quality Award (Malcolm Baldrige) specifically require cross training. It might be noted that much of the content of the Malcolm Baldrige Award is contained in *The Just-In-Time Self Test*. They are very similar and focus heavily on the same total quality management concepts that a good JIT implementation contains:

> Mobility, flexibility, and retraining in job assignments to support employee development and/or to accommodate changes in technology, improved productivity, or changes in work processes.

To meet this requirement for an effective quality organization, an employee development plan should be in place that offers varying job assignments, cross training, and special projects. There should be individual plans for each employee. A typical plan would list specific training that should be accomplished over future periods. It should list what the employee needs to develop his or her current and projected job or special project assignments, along with a schedule for completion of these activities.

A good set of guidelines for assessing your employee development plan would be:

1. A written plan for employee development and cross training.
2. All levels of the company included by specific person.
3. Detailed knowledge and skills needed by employee.
4. The employee's previous experiences, departments and skills.
5. A list of jobs or certifications each employee had done.

Cross-training is nothing more than portable capacity. Imagine the benefit to a company if all people were capable of doing everyone else's job. A bit ideal, but a goal to shoot for. A cross-training program will pay for itself in flexible capacity. It must be planned, tracked, and reported as a continuous measure of the training department and the management team. A good method for tracking is found in some software systems designed for training departments. A manual tracking tool can be set up on a computer spreadsheet similar to the following example.

Cross-Training Matrix

	Insp	Ass'y	Test	Insert	Jig	Final	Total
John W.	X	X	X				3
Roland K.							0
Dan R.	X	X	X	X	X	X	6
Lee D.	X	X					2
Robert B.	X		X	X	X	X	5
Mary W.			X				1
Jack C.	X		X	X			3
James C.			X		X		2
Dan S.	X				X		2
Richard L.	X	X	X		X	X	5
Jeri J.	X		X		X		3
Tom T.	X						1
Total	9	4	8	3	6	4	

It is obvious that Dan R., Robert B., and Richard L. are the most cross-trained, and that Roland K. and Mary W. need a lot of training. Notice there are few people cross-trained to work in the "Jig" department and that many have been trained in inspection. This type of matrix can be very beneficial for both decision making and development tracking.

In a small company this chart may represent everyone on the payroll. This gives all the more reason to focus on cross training. As the company grows, it would be wise to keep this strategy in place, thus building a highly flexible workforce from the ground up.

This example shows the number of employees trained or certified, or both, in six areas. It highlights the fact that "assembly" and "final" have only three people available to do the work in those areas. On the right side of the matrix the employees who are most trained are highlighted. A number of decisions can be made from the matrix, like who should be trained next, where you're short of skills, who is most skilled, and so on.

> Question 6. Compensation and rewards for employees are based on both employee flexibility and team contribution.

Rewards should be given to groups based on team contribution. Data on quality, throughput, cost, and timeliness could be collected on group performance. Group performance is less threatening, is more team-oriented, and operates under peer pressure. There must be a set of clearcut company objectives in terms of quality production output. The teams can then choose how they will meet those objectives.

Compensation usually is based on seniority, level, or job. This type of system will inhibit quality performance because it is not based on performance. A large percentage of all employees' pay should be based on individual and group performance to quality goals and standards. Too often bonuses are only given to management, not to labor. Ask these questions when evaluating your employee recognition and compensation policies and practices:

1. Is a large portion of the employee's pay based on performance to quality, throughput, or timeliness goals?
2. Are all categories and levels of employees participating in goal-oriented quality-based compensation programs?
3. Are the number and amount of bonuses and special recognitions sufficient to motivate employees?
4. Do the employees think the rewards are commensurate with their effort.

There are a number of cloudy areas here. If a union is present in the plant, compensation will be difficult to adjust based on individual or group performance. Also, many times the operator is constrained by the process he or she works within. However, compensation should be used as much as possible to reward positive behavior. But it isn't easy.

> Question 7. Department or team problem-solving groups meet regularly to address and solve quality

and flow problems throughout all departments.

Problem-solving groups are the very nature of successful JIT. Every work area of the company can be a laboratory for experimentation and improvement. Many of the flow problems are able to be resolved at the work areas. Everyone in the company should be a member of problem-solving groups. Some groups can be organized by work area. Some can be organized by department, and some by special project. Some employees may end up on more than one team. The grass-roots solutions that are developed by these teams are often faster and more easily implemented.

> Question 8. All associated functional departments (e.g., engineering, purchasing, marketing, and accounting) are part of problem-resolution teams.

The first area in which problem-solving groups are formed is usually production/operations. The last place they form is in the support groups. This is exactly backwards. In today's top notch manufacturers, the customer-supplier flow begins long before production. It begins at the first step in the process, from the original external customer all the way through to the shipping department.

It must be noted here that the JIT principles of waste reduction, cycle time analysis, setup reduction, flow charting, pull systems, signaling, and lot sizing all apply to every organization. These principles or techniques are just more visible in production. A manufacturer on the West Coast recently redefined its monthly and quarterly awards system to accommodate a special award structure for support groups like accounting, engineering, plant maintenance, information systems, personnel, and training. These awards were based on exactly the same criteria as manufacturing: increase throughput, increase quality, increase customer satisfaction, reduce operating expense, upgrade the processes, reduce waste (inventory), reduce space and other measures. The firm's work groups are turning in remarkable continuous improvement results.

Upstream departments are typically the major *root* causes of production failures. Dr. Deming says: "Many Americans are trying to *start* with Just-In-Time, unaware that this process is years off. Just-In-Time is focused downstream. It's a natural occurrence." Every company in America has had problems with differing goals between groups. Problems are usually a result of conflicting or unclear goals. Purchasing departments, for example, place orders without understanding the use of the materials, and then, when they fail, purchasing is held responsible. Design engineering often designs products without much knowledge of how the manufacturing process works. Then, masses of after-the-fact engineering change orders are issued. These and other departments are doing what they feel is right but are not tied to the same objectives or had the same training.

One of the best ways to get everyone focused on the same objectives is to have them participate in problem-solving teams either within or external to their departments. The root causes of the majority of problems to be resolved usually require interfacing with other departments. This creates cooperation and raises questions about goals that might be conflicting. To facilitate a better coordination of goals and objectives, Safetran Corporation, an electronics assembly plant, organized a total quality project team to facilitate groups in all areas of the company. The first task was to take all of the ideas generated by the local department teams and prioritize them based on the following criteria:

1. Does this idea increase throughput to sales, not to some inventory stocking location, while simultaneously . . . ?
2. Does this idea reduce inventory or the cost of carrying inventory, while simultaneously . . . ?
3. Does this idea reduce the operating expense that it takes to convert inventory into throughput?

Once prioritized, the list of ideas was sent back to the teams with the suggestion that they might look first at the top five items on the list as candidates for their problem-solving activities. The teams were then facilitated by each of the team "sponsors" on the TQM project team. Some sponsors had more than one problem-solving team to support. Any team could call any other team member to its meeting for help in solving its problems. This

FIGURE 5-1
Project Team Organization

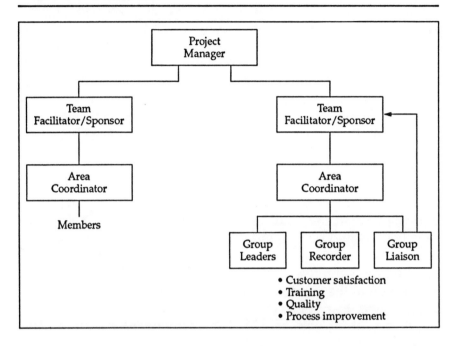

worked well in building inter-departmental communication. As the teams became used to sorting by the priority scheme set by the TQM committee, they evolved into their own priority list and continued with the support of their sponsor and local supervision.

The problem-solving groups, called TAC teams (team action concept teams) were organized in a manner to facilitate better communications across departments also, as shown in Figure 5-1.

The TAC teams at Safetran were so structured that each team had a group coordinator, a group liaison, and a group recorder. The *coordinator* was responsible for planning the meetings, interfacing with the team sponsor, calling the meetings, setting or tracking the agendas, and helping to keep the team on track. The *group liaison* was responsible for making notes and observing the team process, task lists, decisions, and accomplishments. These written observations were then reported to the team sponsor after every meeting so the activities could be reported to the TQM project team for updates. The *group recorder* was responsible for taking minutes and producing them through the team coordinator for the next meeting. The team leaders were broken into four groups: (1) customer satisfaction; (2) training; (3) quality; (4) process improvement. Each of these subgroups was charged with the responsibility of problem identification and resolution in each of its areas. The subgroups were asked to plan, do, check, and act on any problem they selected that met the original prioritizing rules set by the TQM committee. There were two 30-minute meetings held each week, which eventually evolved to a one-hour meeting each week.

Once every quarter each team was reviewed by a select team of employees from all areas of the company. These were shop workers, supervisors, engineers, the president, accounting, purchasing, and the like on the quarterly quality review team (QQR). The members rotated every quarter. They had a series of questions to ask for all groups, which they generated themselves. Each TAC team had to submit a written report of all improvement projects worked on in the quarter. It was reviewed in advance by the QQR team. Great learning occurred to all members. As many as 40 improvement projects were implemented over each quarter. This quality review process was the absolute heartbeat and perpetuator of the TQM process in the company.

This example is a reasonably simple model to follow. The key to getting all the departments involved, however, still rests with supervision, who must require all departments to conduct problem-solving groups in their areas. Continued effectiveness of team problem solving is maintained by how well the management group handles question 9.

Smaller companies can do the same things as defined here even if they only have one or two teams. The smaller the company, the more intimate the team. The exercise of problem solving will help formalize the usually informal nature of their meetings and focus them on more professional conduct.

> Question 9. Management responds to employee ideas and feeds back the manager's responses immediately.

Historical research shows that problem-solving teams (i.e., quality circles, etc.) will fail if no feedback about their efforts is provided from higher-level management. The key to success of these two-way communication channels is the project manager. This carefully selected person can prevent the confusion caused by sporadically reported activities. By organizing the flow of all team activities and, in conjunction with his or her team, the project manager acts as a clearing house that focuses and reports team performances to the steering team or chief executive officer (CEO). The quarterly quality review process also helps.

The most common action required from a CEO or steering team is the definition of policy. Policy drives procedures that result in measurements of performance. Team performance is the result of policy making and the monitoring of performance. Chapter Twelve defines the proper mechanism for establishing and monitoring world class measures necessary for effective Just-In-Time implementation.

Just-In-Time Action Plan—Education and People

List those items you discovered in this chapter that should be added to your JIT Action Plan.

1. _____
2. _____
3. _____
4. _____
5. _____
6. _____
7. _____
8. _____
9. _____
10. _____
11. _____
12. _____
13. _____
14. _____
15. _____

Chapter Six

Quality Management

Self-test users must be aware of the importance and impact of quality management on the Just-In-Time systems they intend to implement. There is no doubt that quality in all activities is critical to the success of JIT. As you read through this chapter, you will be shown how to build quality in all aspects of the business and how to test for the effects of overall quality failure. Examples and references will be provided for you to review. At the completion of this chapter, you will have reviewed or learned the following quality concepts and applications:

1. Total quality management (TQM) and associated concepts.
2. How to build quality management into the strategic business plan.
3. What ISO 9000 is and how it can effect you.
4. How to set up and implement a supplier quality certification program.
5. The cost of quality and how to track and report it.
6. The impact of process variability and human nature.
7. How to use and apply statistical process controls (SPC).
8. How to eliminate incoming and in-process inspection.

Don't forget to keep a list of the detailed items in this chapter that you find missing in your own operations and add them to your JIT Action Plan. Check the Bibliography for other readings that will directly apply to your plans for JIT. This chapter is probably the most important one in the book because it is the *quality* chapter. Unfortunately, it cannot cover all of the necessary tools and techniques to implement a full quality process in your organization. The Bibliography contains some excellent books on total quality management and on statistical quality control. This is one area in which *everyone* must be involved.

High-quality work in all areas of operation and management substantially enhances and facilitates a JIT implementation. *Variability* in management systems and production processes often creates a need for excess buffer stocks. As quality improves, excess stock will be eliminated. Keep this rule in mind: "All management errors end up in inventory!"

> Question 10. "Management exhibits consistent support for quality procedures."

Top management must plan for the implementation of quality systems in all areas of the organization, not just in production or quality assurance. Large amounts of work in every department every day are done poorly: schedules are not correct, inventory is not accurate, engineering changes have errors, orders are not entered correctly, design engineering leaves out parts, purchase orders list the wrong parts, production control issues the wrong orders, data are lost by information systems management, and much more. All of these events can and do create quality problems throughout the logistical flow from your suppliers to your customer. There are many strategies that can be outlined to overcome these failures, total quality management being one of them. Note that as much as one-third or more of the cost of most operations is hidden in the realm of poor quality. Concern about waste reduction is needed from everyone.

Total Quality Management (TQM), like Just-In-Time, is a never-ending journey. The central concept is continuous improvement. A simple definition is: *Doing it right the first time—every time.* A broader definition is: *Any characteristic of any product or service that makes it less than totally desirable for use.* An even better definition is: *Getting people to do better what they should have been doing anyway!* The technical definition of TQM is: To provide a product and service into which quality is designed, planned, marketed, produced, and maintained at the most economic cost, which allows for full customer satisfaction. However, many times the operator or support employee cannot control the process in his or her area. It may be controlled by others or by incorrect policy. These things become issues to resolve in problem solving for continuous improvement.

FIGURE 6-1
Strategic Quality Planning

Cost Strategy	Cycle Reduced	Customer Response	Internal Resources	Planned Contribution
	X	X		Higher-quality output in new product changes
X	X		X	Reduction of overall cost of quality in all departments
X	X	X		Increased quality competitiveness
X			X	Reduction of quality inspection staffing
		X		Implementation of quality indicators to customers
			X	Employee improvement activities to improve work structuring
X		X	X	Reduction of field support quality costs
X			X	Improved management of operations coordination of quality
		X		Improved visibility of quality systems
	X	X		Improvement of product design, modification, and performance

All these definitions lead to the conclusion that the company must have an overall philosophy of quality. No longer can any department, group, or person act as an individual making unilateral decisions that optimize their own interests at the expense of others. This is the essence of quality management—there must be a quality strategy that begins with the business plan and flows downward and outward throughout the organization. Figure 6-1 is a quality strategy example that should be contained in a strategic business plan.

A number of observable positive impacts are derived from a

strategic quality plan. The columns to the right in Figure 6–1 indicate the impact on the company's cost reduction strategy, overall cycle time reduction, positive customer response to better quality and delivery, and the better employment of internal resources as reflected in productivity or return on assets employed.

The key to success in any venture is, of course, having a plan. The plan is driven by a strategy. It is the responsibility of top management to define a quality strategy and then to support all subordinate activities with consistent decision making.

A quality strategy surrounds all other activities. TQM surrounds the JIT systems, which surround the processes and the changes that will occur to them as JIT is implemented. Inside the processes is a quality monitoring system (statistical process control —SPC). SPC identifies what quality changes to make to the processes to meet the JIT goals, which ultimately fit the quality management strategy defined in the business plan.

Once the plan is in place, a tracking and reporting mechanism needs to be implemented to identify the four major elements of the *cost of quality* to the organization: prevention, appraisal, internal failure, and external failure.

Figure 6–2 illustrates how the costs of quality can be tracked and reported. Each department reports any quality-oriented costs to a general ledger account number. The accounts then are totaled and segmented by type of cost, by department.

The objective of this type of reporting is to reduce or eliminate the bottom two elements of cost—internal and external failures—and to redistribute some of that cost to the top of the chart—prevention. It is easy to see that costs like scrap, rework, and warranties would be eliminated as error prevention is increased, thus allowing greater funds to support prevention.

Another approach might be to consider prevention and appraisal as *quality investments* and internal and external failure as *costs*. Then it is possible to show how increased investment yields benefits in cost reduction.

Question 11. "Inspection sequences have mostly been eliminated and quality is part of the individual operator's responsibility."

FIGURE 6–2
Total Cost of Quality

Acct. No.	Type of Quality Cost	Dept. 202	Dept. 203	Dept. 204	Totals
1.01	Quality management		$10,311	$ 28,734	$ 39,045
1.02	Process studies				
1.03	Quality information				
1.04	Training				
1.05	Misc.				
1.0	Prevention subtotal		$10,311	$ 28,734	$ 39,045
2.01	Incoming inspection			$ 4,568	$ 4,568
2.02	Calibration and maintenance	$ 2,937	$16,717		$ 2,937
2.03	Production tests	$52,256			$ 68,973
2.04	Special tests and audits				
2.0	Appraisal subtotal	$55,193	$16,717	$ 4,568	$ 76,478
3.01	Scrap			$ 55,752	$ 55,752
3.02	Rework—producion	$ 7,410	$ 4,869		$ 12,279
3.03	Rework—vendor failure	$ 246			$ 246
3.04	Corrective action	$ 3,069	$ 2,630		$ 5,699
3.0	Internal failure subtotal	$10,725	$ 7,499	$ 55,752	$ 73,976
4.01	Warrantee expenses	$ 2,706	$12,106		$ 14,812
4.02	Postwarrantee expenses				
4.03	Customer services			$ 52,765	$ 52,765
4.0	External failure subtotal	$ 2,706	$12,106	$ 52,765	$ 67,577
Total		$68,624	$46,633	$141,819	$257,076

Many manufacturing routings contain "inspection" steps. Inspection by its very nature implies quality failure. It exists based on the false assumption that quality failure cannot be detected or controlled by the operators. There is a tacit agreement that it is OK to have some failure. Some call it an "AQL"—acceptable quality level. Thus, the purpose of inspection built right into the process is to appraise or assess the acceptable quality failure levels.

Sometimes, unfortunately, the failure levels that are "accepted" are even higher than many competitors. As a result, warranty repairs, replacements, service costs, and other costly consequences occur. The elimination of inspection sequences from the processes creates a considerable change in company philosophy.

The best method to begin eliminating inspection from the processes is to measure the process or the operator's performance. There are a number of tools listed in Chapter Five—questions 7, 8, and 9—about people involvement that may be used to start a quality awareness program, specifically having small meetings in the work areas to address quality problems. As these activities increase and become more effective, quality tracking and reporting will increase. As the problems are removed, the need for inspection will be reduced and eventually will be eliminated. As the inspectors are eliminated, they can become trainers, process auditors, facilitators, or just operators. The operators should make their own quality measures and receive feedback first hand. There should be a concentration on fail safing the process to prevent errors. These are the results of team problem solving. It's easy to see the value of good teams.

> Question 12. "Quality control departments have been replaced with process audit and operator training functions."

If the quality control or quality assurance departments are removed from the process sequences, what else can they do that will utilize their skills and talents? How will the quality be known? There are two areas where they can be more qualitatively utilized—process audits and operator training. The ultimate extension of a quality inspector is to feed signals back to the operator to identify defects. Although the inspector may not have the skills of the operator, he or she has the skills of analysis and evaluation. These skills can be shared. Minischools can be established and run by quality personnel. Suppliers as well as internal operators can be taught the techniques and disciplines of statistical process control.

Process audits are not random inspections to check if the operators are still making good parts. Process audits are done on the whole process itself. Using the charting techniques defined in question 13, the former quality inspector's role is to chart and analyze variable processes, looking for quality trends. They will interface with engineering, manufacturing, purchasing, marketing, and others in search of solutions to process variability.

> Question 13. "Early warning statistical quality control tools are in place and are used to monitor and control quality at critical points in the process."

Process variability is another word for potential quality failure. The more variable the process, the greater the quality failure. In all process there are greater and lesser amounts of variability. The most critical ones are those that will cause the greatest rework, MRB (material review board, a quality defect decision activity), or shutdowns.

An early warning system for identifying and correcting errors and defects is required to prevent the losses of time and throughput caused by quality failure. The most notable early warning system is statistical process control (SPC). It signals a process that is getting out of control or even out of specification at the time it's occurring. Other systems are process audits, which check the entire quality process within a department or group, and fail safe devices, which actually prevent failure before the errors ever occur. These techniques will be explored.

Several proven problem-solving techniques can be used:

a. Process flow charts.

b. Check sheets.

c. Pareto analysis.

d. Fishbone or cause and effect diagrams.

e. Scatter diagrams.

f. Process control charts.

g. Run diagrams.

Process Flow Charts. Process flow charts track the flow of a product through all the steps it takes to make it. The charts are used when you want to identify problems or to find the ideal path to flow a product or process. Each step can be a source of failure, a time-consumer, or bottleneck,—or simply an unnecessary step.

Flow charting is a necessary first step that must be done prior to any process or flow changes. The relationships between each step can be graphically seen, and they often show loopholes or redundancy. By recording times for each step, bottlenecks and idle times can be identified. Cycle times as defined in Chapter Nine can be analyzed to find which bottlenecks can be reduced or where to put extra labor or machinery to increase flow. Flow charts can be applied to anything, from the movement of materials through production to the stages in processing a purchase order or the steps to process and invoice for payment.

Typically, two charts are required, one of the current processes and one of the proposed processes that incorporate the changes discovered from the first chart. Figure 6-3 shows an example of a flow chart.

Note that the beginning and end are ovals; that every process box or rectangle element has at least one "in" and one "out," that every decision diamond has two "outs"—one for each decision. These are simple symbols and can be used in most of the processes analyzed. More sophisticated tools can be used but are not covered in this book.

Check Sheets. Check sheets are a logical starting point, typically used as a preliminary data gathering tool for problem-solving. Their objective is to highlight problems seen from sample observations. The example in Figure 6-4 shows the pattern of defects in fasteners.

Two things are immediately noticeable from this check sheet. First, the wrong size is the most common occurrence, and second, on the 14th there were at least three times as many errors than on any other day. The results of this check sheet can lead to the use of several other problem-solving tools.

Pareto Analysis. Check sheets lead easily to the use of Pareto charts, which are typically vertical bar graphs used to help determine the relative importance of problems and to highlight the biggest or most important problems to be solved first. Pareto charts can be used for continuous tracking of processes to monitor

FIGURE 6–3
Flow Charting Sample

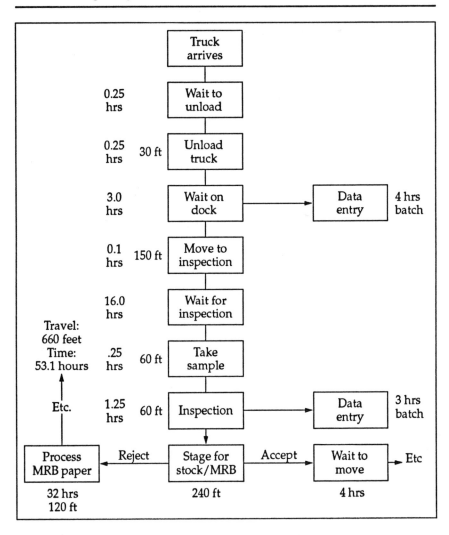

success of changes and to highlight further problems. This simple procedure is often referred to as the "80/20 principle or rule," where 80 percent of failures typically occur in 20 percent of the parts being processed.

To construct a Pareto chart (refer to the chart in Figure 6–5):

FIGURE 6–4

Problem	July							Total
	11th	12th	13th	14th	15th	16th	17th	
Wrong size	////	//	/	면////	//	////	//	21
Wrong length	//	/	/	////		/		9
Wrong hardness			//	//////	//	/	/	12
Inserted wrong	/		/	////	/	/	///	11
Missing	/	///	/	//////	/	//	/	15
Totals	8	6	6	26	6	9	7	68

Step 1: Select the problems that are to be compared. In this example, there are five problems: wrong size, length, and hardness; inserted wrong; or missing. Choose some or all items to chart. Another approach would be to brainstorm as a group the items to track. This might create a larger number of elements to chart.

Step 2: Identify how the elements will be compared—by unit, by cost, by frequency, by period, and so on. In this example, you might choose "number of failures" as the comparison measure, and the days the work occurred within.

Step 3: Gather the data and plot them on a chart comparing the number of failures per period.

Step 4: Sort the bars from highest to lowest. The chart will then show the greatest failures per period. See Figure 6–5.

Fishbone or Cause and Effect Diagrams. Check sheets and Pareto charts lead to the use of cause and effect or "fishbone" diagrams. Pareto charts indicate a problem exists. Now we need a way to determine the possible causes of the problem. The fishbone chart, as seen in Figure 6–6, is used to show the relationship between some effect and all the possible causes that might influence it. In Figure 6–6, the "effect" was cracking, and all of the possible "causes" are charted as shown. For every effect, there may be several major categories of causes. As a starting point, use the typical four categories shown in Figure 6–6: manpower, methods, materials, and machines.

FIGURE 6–5

Number of Failures	Size	Missing	Hardness	Insertion	Length
25					
20	X				
15	X	X			
10	X	X	X	X	X
5	X	X	X	X	X
0	X	X	X	X	X

Again, as in Pareto analysis, brainstorming can be used to develop possible causes without previous preparation. Also ask each person involved in the process to use check sheets and observation to find as many possible causes. Then construct the chart as shown in Figure 6–6. For each cause listed, ask "Why does this happen?" and list the responses as branches off the major causes.

Interpretation of the drawing typically is done by consensus. Ask the group to rate all the causes and branch causes. Collect and sort the responses for the most common cause. A good approach for doing this is the nominal group technique (NGT), which gives everyone an equal voice in the selection of the most probable causes.

Nominal group technique can be used to select which cause to work on and to establish the order or priority of attack. First, have everyone in the group contribute to the list of problems, either from the fishbone, check sheets, or brainstorming. Write the most critical problems on a board or flipchart and code them A, B, C, and so on up to no more than six to eight items. Then have each member make a vertical list of each letter representing each critical problem on a blank sheet of paper, such as the example shown here:

```
Suppliers     A. _____
Rework        B. _____
Scrap         C. _____
Returns       D. _____
MRB           E. _____
Engineering   F. _____
```

FIGURE 6–6
Cause and Effect Diagrams—"Fishbone" Diagram ("Ishikawa Diagram—Dispersion of Causes")

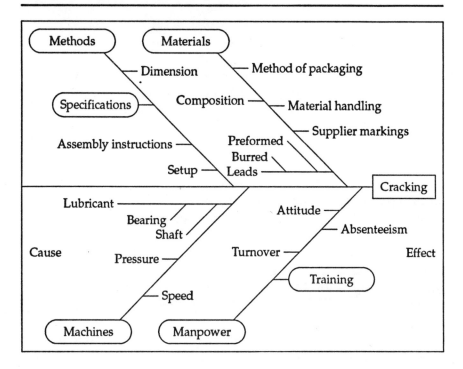

Then, the members must prioritize the list from the greatest problem to the smallest problem. In the above example, one member might select rework as the biggest problem; and, because there are six items in the list, the member would put a 6 on line B, then proceed to rate in declining order the rest of the items, as in the following example:

Suppliers	A...	3
Rework	B...	6
Scrap	C...	4
Returns	D...	1
MRB	E...	2
Engineering	F...	5

The papers are then collected, and all of the ratings are collated and posted on the board or flipchart and totaled as in the following example taken from a group of six people:

			Totals
Suppliers	A...	3,2,4,5,3,2	19
Rework	B...	6,4,3,4,4,3	24
Scrap	C...	4,6,6,3,5,4	28
Returns	D...	1,1,2,2,1,1	8
MRB	E...	2,3,1,1,2,6	15
Engineering	F...	5,5,5,6,6,5	32

The highest total is the consensus of the group. This example shows that engineering is the biggest problem, or probable cause. Once this problem is attacked, then the second highest item (scrap) is selected for action.

This technique also helps to prevent those who might be most vocal or autocratic in the group from forcing the team to work on items they think are the correct priorities without considering the team's input.

Scatter Diagrams. Scatter diagrams can be used to test for possible cause and effect relationships discovered from the fishbone diagrams or other brainstorming activities. Scatter diagrams are used to study the relationship between two variables. As shown in Figure 6-7, one variable is plotted on the horizontal axis and the other is plotted on the vertical axis. The diagram cannot prove that one variable causes the other, but it can show how strong that relationship is.

Notice in the "loose" chart how far apart the points are from each other in a random pattern. In the "tight" chart, where stainless is used instead of aluminum, there is a much greater relationship between the number of items scrapped and the temperature, obviously indicating that aluminum might be a better metal for this process. In the "loose" chart, there is no positive or negative correlation between the amount of aluminum scrap and the temperature. A negative relationship is as important as a positive relationship. In the "tight" chart, the pattern would be trending down

FIGURE 6–7
Scatter Diagrams

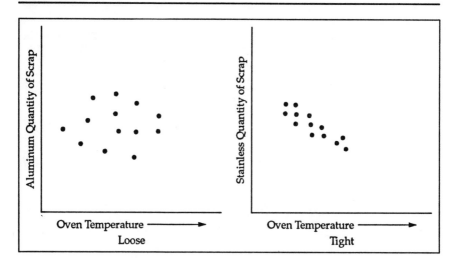

as the temperature increases, possibly indicating that the stainless would be a better metal at higher temperatures. Other statistical correlations and tests can also be used.

Process Control Charts. Process control charts are used after the causes of problems are investigated and some amount of undiscovered or unnoticed variability is found in the process. There are typically two types of causes of this variability, common causes and special causes. Special causes are those looked for first, because they usually occur in unpredictable ways and readily cause failure. Common causes are notable variations that normally occur during the process and that might be refined after the special causes are corrected.

Process control charts identify and plot trends or defects from both special and common causes. Essentially, these control charts act as an early warning system that flags defects when or immediately after they occur, instead of several operations later when considerable more labor and associated costs are added. Catching the defects at the source (question 14) will dramatically reduce costly rework, late shipments, quality assurance costs, and even excess

FIGURE 6-8
Control Charts

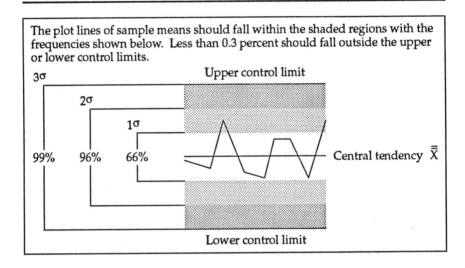

inventories.

A control chart is composed of a central line or average of historical readings along which measured data are posted. It has an upper and lower limit that is generated by a reasonably simple statistical calculation. It also has a "range" chart attached, which measures the widest swing the data sets exhibit at the time measurements are taken. Figure 6–8 shows the characteristics of a process control chart.

Note the line of central tendency with the data or "sample means" plotted along it. Sample means are nothing more than a set of averaged readings taken at each point in time. In this example, nine points are charted. At the time when each point was plotted, typically a set of five parts, readings, or actions were measured or tested. The five were then totaled, averaged, and the resultant average posted to the chart. This action was repeated for each of the nine points.

It also should be noted there are two limit lines, an upper control limit (UCL) and a lower control limit (LCL). These limits are determined by allowing the process to run untouched, within specification, for some time (typically long enough to get 24 sets of four or five measurements). Then a "grand average" or "x double bar" of the averages of all of the sets are calculated. Once the grand average is calculated, then the ranges for each of the sets of five is determined. The range or R of each set is obtained by finding the difference between the highest and the lowest reading in each set. Then an average of the ranges or R bar for all the sets is calculated. With these two numbers, "X double bar" and "R," the upper and lower control limits can be calculated as shown in Figure 6–9.

The following are the formulas used to calculate the upper control limit and lower control limit:

Upper control limit = X double bar + A × R bar
Lower control limit = X double bar − A × R bar

	A_2	D_3	D_4
Subgroup lot size factors 4	0.73	0	2.28
5	0.58	0	2.11

Example: UCL = 8.6 + (0.73 × 2.36) = 10.3
LCL = 8.6 − (0.73 × 2.36) = 6.9

These formulas use the sample group size chart to determine the value of A_2, D_3, and D_4. These numbers (whose derivation is outside the scope of this book), are statistically generated to compensate for the sample lot sizes and are very easy to use: simply insert them into the formula where indicated. The smaller the sample, the higher the probability for variance, so the A_2, D_3, and D_4 numbers help offset the variances introduced just because of the sample size and not the process you are measuring.

The results of the calculations from these formulas are the upper and lower control limits or limits beyond which unacceptable quality variations might occur during the process. To keep it simple, the upper and lower control limits are the lines on the edge of the highway. If you are going down the middle of the road, staying right on the center line and you lose control of the car for a moment, the edge of the road is a signal or limit telling you that the car is about ready to go over the side if you don't steer it back to the center.

Now that the UCL and LCL are done, draw them and the X double bar lines on a new chart called the "X Bar" charts (see Addendum at the end of the book). Do the same process with the range data. Using the formulas for range chart UCL and LCL, insert the average of all the ranges and the group size chart data to calculate the UCL and LCL. Draw these lines on the range chart called the "R" chart (see Addendum).

Note that two of the points (3 and 9) in Figure 6–8 were in the first shaded area above the central line. The shaded areas are statistically generated as standard deviations from the central line or mean. These areas can be determined by simply dividing the distance between the X double bar line into three equally spaced segments. (Similar to calculating "Standard Deviation," see Bibliography.)

You will note each shaded area in the chart in Figure 6–8 has a percent bracket that identifies the percent of probable readings that would normally fall into each area. The farther away from the X double bar line they go, the lower the probability that a reading will fall in them. For example, only 3.7 percent of readings should fall within the third range. These readings are too far away from the center line, and, if more than one or two readings fall here, there is some problem with the process that is being measured. It means the process may be about ready to go out of control. The following are some rules that help define when a process should be stopped or checked. When:

a. 2 points out of 3 are in the third zone (3.7 percent).
b. 4 out of 5 points on the same side of the center line.
c. 6 consecutive points moving up or down.
d. 15 points in the first zone continuously above or below the center.

There are many circumstances where manufacturing processes can get out of control such as gradual tool wear, change of operators, change of raw material, poor training, and so on. Figure 6–10 shows the types of patterns that can occur.

There are a number of more precise statistical tools than this simple approach that include standard deviations, mean absolute deviations, and so on, but those techniques are outside the scope

Chapter 6/Quality Management

FIGURE 6–10
SPC Control Charts Trending

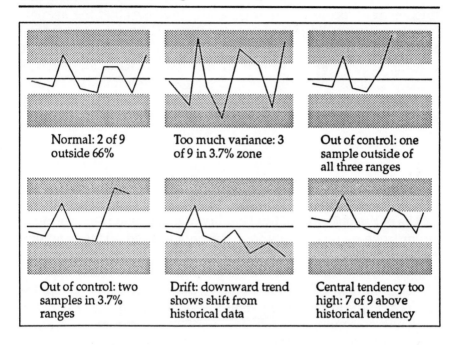

of this book. A more thorough knowledge is required to implement the use of statistical process control than can be presented here. These things should be included in the training plan discussed in Chapter Five.

Process control charts tells us when something in the process has changed. It is a measure of averages of samples. A run chart or diagram might be used where simpler needs exist.

Run Charts. A run chart is a chart of *individual* measurements and is used when a simple display of trends within upper and lower specification limits is needed. A chart is constructed by defining a desired center line between an upper and lower predefined set of specifications or tolerances. Points are plotted on the graph when each piece of data is taken. It works well for processes like scrap, downtime, cash, labor, deliveries, productivity, errors, and returns. Run charts show every variation of the process and should be used to focus attention on truly valid changes.

When between six and nine points fall above or below the target line, or trend up or down, it generally indicates that the average has changed—calling for an investigation. There are more sophisticated tools than the seven basic statistical tools, which may be considered later in the growth of the quality disciplines such as quality function development, design of experiments, Taguchi methods, and failure mode and effects analysis. Unfortunately, these are outside the scope of this book.

> Question 14. "Quality errors are repaired or prevented at the source where they occur."

The entire objective of total quality management is to continually improve quality throughout the organization. This is accomplished by focusing on the source of the errors that generate defects. Defects are waste—a waste of time and of resources. The closer to the source of a failure that a defect can be prevented, the less cost to the organization. This means that products or paper or data will cost more to repair if they move away from the originator. When a quantity error occurs in a sales order, everyone downstream is effected by it. When a defective product ships from a supplier and goes all the way through to production, the cost of failure is very high if a product or assembly has to be reworked or scrapped. Are you fully aware of the impact on your customer when they receive bad products from you? Quality costs must be avoided at the source. Questions 13, 15, and 18 are related here.

This means that rework or unplanned scrap of any kind is an indicator of a quality problem. Rework must be reduced and eliminated. (Some companies have literally institutionalized rework by structuring it right into their process routers.) When defects are detected, an immediate feedback loop must signal to the source of the defect that a bad part or process has occurred. Repairs need to be brought to the originators of the defect to help them realize their impact on downstream operations. Support functions like manufacturing engineering should be immediately available to help resolve problems. Question 15 is a world class technique for preventing errors even before they occur.

Question 15. "Fail-safe (poka-yoke) devices are installed at most locations where recurring human quality errors typically occur."

People are human. They are forgetful and make mistakes. We blame them for failures, we train them to prevent failures, we measure them to catch their failures, we motivate them to do better-quality work, and we even fire them for producing failures. Yet, we still have quality failures. There must be a better way to get to the goal of zero defects. Besides, people get demoralized when they can never seem to produce zero defects all of the time and are constantly criticized.

Fail-safe or "poka-yoke" devices are the solution to this problem. Fail-safing is the highest "inspection" system of all. The error that causes the defect that causes the quality failure is prevented. What better system than this? By using a method for "fool-proofing" a person's work, defects can be prevented by literally preventing the human error before it causes a defect. The concept of fool-proofing has been around for a long time, and many organizations use them already. The idea behind fool-proofing is to use simple, inexpensive devices to take over the repetitive tasks or actions that often depend on a worker's memory or vigilance.

Shigeo Shingo in his book *"Zero Quality Control: Source Inspection and the Poka-Yoke System"* describes many examples of fail-safe devices. His focus is to use the workers to generate ideas for the devices by simply asking them to find ways to fool-proof their work flow. A most common example of a fail-safe device is the standard wall plug. It has three holes oriented to fit all standardized plugs. The plug can never be put in wrong. This way, people won't get electrocuted because they hooked the live wire to the ground wire by plugging it in wrong.

The following list of errors caused by workers are examples of the kinds of things that fail-safing will be able to prevent:

KINDS OF ERRORS CAUSED BY WORKERS

Unfortunately, there seems to be no shortage.

- *Forgetfulness*. This can be eliminated by alerting the operator

in advance or checking at regular intervals.
- *Errors due to misunderstandings.* Safeguards for this include training and re-training, checking in advance, standardizing work procedures, pictorial instructions, and so on.
- *Errors in identification.* These can be reduced by training, simple templates, attentiveness, vigilance, and in some cases by electronic vision inspection devices (pattern recognition).
- *Errors made by amateurs.* We can go a long way toward eliminating these by skill building, work standardization, and effective first-line supervision.
- *Willful errors.* We cannot allow people to "get by" without doing what is needed. Rules cannot be ignored. Give them basic education and experience and enforce the correct method.
- *Inadvertent errors.* This turns out to be the largest cause of human error. Absentmindedness and making mistakes without knowing it are real problems. We can eliminate these by attentiveness, discipline, work standardization, and poka-yoke devices.
- *Errors due to slowness.* We can speed up our people's ability to make effective correct decisions by skill building, work standardization, training, and practice.
- *Errors due to lack of standards.* We can drastically improve this by providing pictorial work instructions or models of correct parts, and by work standardization.
- *Surprise errors.* Stopping machines from malfunctioning without warning requires total productive maintenance and work standardization; safety devices, such as limit switches, may help.

The idea behind poka-yoke is to respect the intelligence of the workers and yet to make the production process stimulating. By taking over repetitive tasks or actions that depend on vigilance or memory, poka-yoke devices can free the worker's time and his or her brain to pursue more creative and value-adding activities.

Poka-yoke is not limited to production aids, or even to manufacturing processes. Often the suggestions of workers create the need for a minor design change to the product itself, to eliminate the possibility for error. Design engineering departments that first allow production to experiment on prototypes, before finalizing a new design, will undoubtedly receive lots of these suggestions. If the name of the game is *throughput*, then they must listen and implement the bulk of these suggestions. Poka-yoke has also been implemented in clerical operations, in sales departments, in documentation storage areas, and in many nonmanufacturing environments.

You do not need to have a highly automated factory to use poka-yoke devices. Most of these devices are extremely simple and many are manual. Most cost less than $50 to implement. The savings generated by poka-yoke devices, in concert with source inspections and immediate corrective action, are incalculable. It is safe to say that *all* world class competitors have implemented many poka-yoke devices as one of the techniques they used to achieve world class.

Defects exist in one of two conditions: either they are about to happen or they have already happened. Poka-yoke devices have three basic functions to use against defects—shutdown, control, and warning. Recognizing that a defect is about to occur is called "prediction," and recognizing that a defect has already occurred is called "detection." For *prediction* events, poka-yoke devices are used to shut down the machine or operation when a defect is predicted; or to narrowly control the operation of the equipment within preset parameters to prevent even intentional errors; or to give a warning to the operator that an abnormality or error is about to occur. For *detection* events, poka-yoke devices are used to stop the operation or even the entire line when a defect is detected; or when using flow control devices, to physically prevent defective items from being passed on to the next process; or to give a clear warning to the operator that a defect has occurred, and that he or she must take corrective action *now*.

EXAMPLES OF POKA-YOKE DEVICES

There are two classes of components used to build poka-yoke devices for detecting errors and defects: those that contact the part being tested and those that do not contact the part. In the category of contact components are the following inexpensive commer-

cially available parts:
- Microswitches.
- Limit switches:
  —Pin push-button type.
  —Panel-mounted roller push-button type.
  —Hinged lever type.
  —Hinged lever with roller type (most common).
  —Hinged lever with roller type operating in one direction only.
  —Roller-leaf spring type.
- Proximity switches.
- Positioning sensors.
- Displacement sensors.
- Metal-passage sensors.
- Transducers.
- Thermistors.

In the category of *noncontact components* are the following inexpensive commercially available parts:

- Photoelectric switches:
  —Transmission type (requires transmitter and receiver unit).
  —Reflecting type (requires one unit, but reflecting work piece).
- Counters.
- Templates.
- Material handling devices.

For example, a poka-yoke device was created by a company that had problems in drilling a certain rectangular-shaped piece. Often the operators would install the workpiece backwards and drill two offset holes in the wrong place. The defects were not discovered until assembly. A limit switch was mounted on the jig to detect grooves cut on two sides of the workpiece. When the workpiece is backward, the limit switch is activated and the machine cannot operate. Defects due to drilling mistakes in this process were completely eliminated, achieving zero defects. This obviously required some assistance from engineering to add the two grooves to the design of the workpiece in an area where they would not affect the performance of the part.

In a packaging operation, line workers had to pack 10 small accessories and an instruction manual in a cardboard carton. Often they forgot to put in certain parts or left out the manual. The cartons often were returned by angry customers. After analyzing the problem, bowed springs were put in each parts container, connected to limit switches. The act of removing a part tripped the limit switch, which lit a green lightbulb. The act of removing the manual was tripped by a photoelectric beam, also connected to a green lightbulb. After the operator had picked all the parts, he looked at all the lightbulbs to see that he hadn't forgotten anything. In a further improvement to guard against intentional errors, the cardboard boxes were set on a roller conveyor with a stopper between the rollers. The stopper would not disengage, thus letting the completed carton roll down the line, until all 11 lights were lit.

By using the tools listed in question 13, the cause or the location of most common defects can be found and poka-yoke devices applied by the workers.

The approach is simple: Inform the workers of the types of devices that have been done in other areas or companies, then ask them to look for applications for more devices. Then make sure you support their suggestions.

> Question 16. "Work areas are consistently clean, organized, and free of unnecessary materials and equipment."

Quality begins with good workplace organization. A clean, organized work area promotes discipline and is the first step to starting up a JIT project. There are several actions to be taken to organize the workplace for quality:

1. Clearing the area. Remove all unnecessary items from the work areas, such as worker's personal items, lunch boxes, and plants. Also remove any excess tools, inventory, or equipment. Store those items elsewhere until needed. These actions will increase visibility of problems, excess failures, bottlenecks, and so

on.

2. Ordering the area. Define where everything goes. All tools, gauges, parts, equipment, incoming and outgoing parts locations—everything must have a place. Let the workers lay out the plan themselves. Label the places where things belong so everyone knows where to put them when they are finished using them. Group items together that are used together.

3. Cleanliness. Clean workplaces show that quality and care exist. Safety also is enhanced by clean work areas. Cleaning equipment is as important as cleaning work spaces. Leaks, overheating, cracks, and other equipment can be readily seen if equipment is kept clean, thus preventing breakdowns. This will also establish a sense of ownership and responsibility for the operator.

> Question 17. "The majority of incoming materials are certified or source inspected at the suppliers."

Supplier quality failure is the reason why receiving inspection exists. It must be recognized that JIT cannot succeed without high-quality incoming materials. The majority of incoming materials can simply bypass receiving inspection if the quality and quantity were certified by the supplier in agreement with your organization. This may look like a large task to accomplish, but remember, 80 percent or more of the volume of parts received probably come from 20 percent or less of the suppliers.

During the supplier certification process defined in question 18, a number of suppliers are selected by using Pareto's principle, or 80–20 rule. The first step is to sort the supplier base by determining which suppliers provide the highest volume of parts flowing into the company. Rank the suppliers in descending order by incoming volume. Select the top few suppliers that represent the largest volume and set them up as candidates for supplier certification. Once these supplier begin getting certified, the volume of parts needing inspection drop.

As mentioned in question 14, the cost of quality failure is greater the farther a product is from its source. This means suppliers that were selected as candidates for certification should be inspected by source inspectors as soon as possible. That way, control is closer to the source and feedback is faster. Source inspectors can be created from the current receiving inspectors. As full certification occurs to those suppliers, source inspection can be eliminated and staffing can be reduced or transferred to other functions.

> Question 18. "Supplier Quality Certification and performance rating programs are in place and continuously monitored."

Three major areas must be done well if JIT is going to work: engineering, production, and purchasing. The longest lead time of the three is purchasing. An excellent way to get Just-In-Time deliveries and zero defect quality is through a supplier certification and partnering program.

It doesn't take long before a supplier certification program becomes very complex and confusing. If you are going implement a supplier certification program, you are going to need a simplified set of rules and tools that will speed up and clarify the certification process. It isn't easy, but the payback is exceptional and worth the effort.

The purpose of supplier certification programs is ultimately to eliminate waste, excess costs, and poor quality that exists between supplier and buyer. Dr. W. Edwards Deming tells us that "quality control should check the process, not the product."

One of the fastest growing trends in the United States is ISO 9000. It is a set of international quality guidelines and audit processes used to assess a supplier's ability to produce a high-quality product. It is based on an internationally agreed set of quality criteria audited and validated by independent quality auditors whose job is to assess a company's quality procedures and processes. The results of these audits are used to highlight internal improvements in quality management as well as to certify to the world that the company produces a quality product. The key to success in an ISO 9000 audit is to be performing the written quality procedures as they are defined.

ISO 9000 and its associated parts are accepted by over 91 countries around the world at this time. More countries are continuing to subscribe. If your company intends to do business with any of those 91 countries, you must be registered as an ISO 9000 organization. Many companies are becoming ISO 9000 certified even if they are not doing business overseas. Some are doing it because their customers are requiring it. Some are doing it to enhance their marketing image of quality. Some are doing it merely to help themselves develop a better quality system.

ISO 9000 is aimed at the heart of the buyer–supplier chain. Many companies that are implementing TQM are aware of the value of the ISO 9000 in identifying their quality failures and improvement opportunities. In essence, customer quality certification programs are the same as ISO 9000. They have the same objective in mind—quality products or services. We may see in the reasonable future a standardized set of supplier certification audits and certificates—by industry—that will eliminate the hundreds of customer audits that are now being done.

So, how do you set up a supplier certification program?

Most of the excess cost, waste, and poor quality is tied to the supplier–buyer chain. So, the first step is to clearly define a companywide objective that might read like this:

> Supplier certification should provide products and services into which quality is designed, planned, produced, delivered and maintained at the most economic cost that allows for full customer satisfaction, increased profits, and mutual benefit for both parties. The results must be mutually beneficial. The key success measurement of this program will be *defect-free* delivery from the supplier's production or stock directly to our production without inspection.

ESTABLISH A CERTIFICATION TEAM

It seems every new program that comes along has to be done with a team. That's because the success of those programs won't happen if done by one person or a disorganized group. The size and impact of a supplier certification program requires commitment from company management, which is vested in a certification team and a certification representative to head the team. The "cert rep," typically a quality engineer, is responsible for acting as a liaison between companies. He or she will lead the certification team, verify specifications, monitor lot inspections during startup, initiate corrective action, and manage follow-up audits. This is a tough job and requires the quality engineer to be a strong manager with excellent product and process knowledge.

A person with these skills will already be valuable to the company and will be difficult to get free to manage this program. Jack Conrad, a division general manager for Safetran Systems Corporation, a maker of railroad crossing controllers, told how he was able to get a key person to head up his supplier certification program:

> Our key process engineer and systems manager was in an unfortunate car accident recently. We had to rely on others to continue his work while he was gone. We found that the work got done—certainly a lot slower than he would have been able to do; but when he came back we were able to shift him to our supplier certification project and take full advantage of his well-rounded skills. We recognized that the true long-term success of our world class manufacturing strategy must include a strong supplier base, and we can only achieve that objective with a well-managed internal certification program.

Each supplier should also have a cert rep who will be the direct link between companies, oversee and audit outgoing shipments, manage corrective actions and reports, and product or process changes to the customer. This is usually a senior inspector or a quality engineer.

Have you noticed that the buyer doesn't seem to be in the loop? Not quite. The rest of the certification team is comprised of members from purchasing, engineering, manufacturing, quality, and planning.

One of the first tasks for the team to do is a cost-benefit analysis to define the certification program savings that comes from the elimination of inspection. The algorithm is simple:

Cost Savings =
Incoming inspection costs + Quality and reliability costs − Certification costs

Incoming inspection costs are a total of *(a)* all receiving inspection labor; *(b)* receiving inspection space; *(c)* reject handling; *(d)* quality engineering support for receiving inspection; *(e)* capital equipment usage for receiving inspection; *(f)* inventory carrying cost for inventory normally in receiving inspection.

Add to these savings the cost of rework, shutdowns from poor-quality parts, longer lead times, higher inventories, and indirect labor associated with all these. This will be big number. Just think about all the waste of time and materials throughout the company just based on bad parts from suppliers. It'll take a bit of research to get these numbers, but they will be big.

Now, subtract the cost of setting up a certification program from the above two totals and that is your justification.

Defining Certification Criteria

Do an ABC analysis of the supplier base by ranking the supplier from the highest to the lowest volume of incoming parts. Select the suppliers that are in the top 10 to 20 percent. Most companies already have some amount of supplier performance raw data in their files—like percent of lots accepted, number of days late, and so on. Use this data to rank the high volume suppliers from high to low quality.

Start your program with the highest-quality suppliers first. Success will be easier and breeds success quicker. A division of 3M recently found that only 10 suppliers represented 40 percent of all incoming parts. Safetran Systems found that 10 suppliers provided over 90 percent of its incoming parts.

Then, rate the selected suppliers against the following criteria:

1. No rejections in the last six months.
2. No negative incidents in the last six months.
3. They have a documented quality system.
4. They will produce timely test and inspection data when required.

It may look like the list of certifiable suppliers is very low compared to the total number of suppliers. But, remember, the suppliers were sorted by the items that require the greatest volume of inspection and handling time. These suppliers will give the greatest payback.

Selecting the Parts to Certify

The biggest problem is to make sure the right parts to be certified are defined for each supplier. You may not want to certify them all. Check each part against the following list:
1. Does the part change a lot? How about the future?
2. How often do we buy this part?
3. What is the average lot size?

You don't want to certify a supplier for parts that change a lot, or that will be declining in volume or purchased infrequently in small lots. These parts will be listed in a supplier contract, which defines the buyer-supplier relationship and the specifications for each part.

Each part must qualify by undergoing (a) a first article inspection; (b) a specification accountability check—a product specification form attached to the contract defines the machine or processes used, the points of quality testing for each characteristic, and the method and frequency of verification; and (c) inspection of three consecutive lots at 100 percent. A random audit should be conducted periodically thereafter.

Defining a Certification Contract

Unlike the normal purchase order, or blanket agreement, the certification contract defines which parts will be certified, how their quality will be checked, and when the supplier's quality systems will be audited.

The objective of the contract is to establish and maintain surveillance and audits to ensure a continuous supply of quality parts direct from the supplier's processes to the buyer's processes with minimum handling and *no* inspection.

Contract management should be guided by the following sample criteria: (1) delivery ratings at or above 99 percent; (2) quality audits done every year; (3) no more than 6 months have passed since the last shipment, and (4) recertification to be done within 24 months.

Scheduling a Kickoff and Supplier Conference

The purpose of this meeting is to define the goals and mutual benefits for all parties. It will define the qualification, survey, and audit process. Open discussion is welcomed. It sets the products to be considered for certification by supplier, the implementation time frames, and procedures for all upcoming activities.

Performing the Survey

Many published surveys can be used to validate the supplier's process. Remember what Dr. Deming said? The survey acts as both a measuring device and as a model for improvement. An overall rating scheme should highlight shortcomings as well as competencies and be reported immediately on completion to the supplier's management. The certification team conducts the survey, evaluates the results, determines which suppliers can be immediately certified, and defines the corrective actions for the others to become certified. Follow-up audits must also be set.

Awarding the Certification

A certification award, usually a plaque, should be awarded to every supplier who qualifies under the program. The award should be made in public at both the buyer's and the supplier's facilities. This is a great supplier relationship tool. A good certified supplier relationship becomes a partnership. Supplier certification is not just a commitment to a quality program, it is a commitment to a quality partnership in which both parties mutually benefit and grow.

A review of the supplier certification process activities that must be done are:

1. Define and staff a certification team.
2. Identify and select the key suppliers that will be part of the program.
3. Rate the current supplier's quality and delivery performance.
4. Determine the products or services that the suppliers

will be providing under the program.
5. Define a contract that lists the products and specifications of the parts selected for the program.
6. Conduct an initial supplier certification kickoff meeting.
7. Conduct the supplier surveys.
8. Evaluate scoring and qualification for the selected suppliers.
9. Present certification award and contracts to the suppliers.
10. Follow up and monitor religiously after the award.

The quality function belongs to everyone in the organization. It begins with top management defining a strategy that focuses on prevention, not an effort to continue the status quo. One of the major objectives of TQM is to eliminate in-process inspection and incoming materials inspection. This can be done by forcing the responsibility for quality back to the source and then giving the source the proper quality tools to work to become successful. Focus should be on error prevention at the source. One of the best known tools for this is the use of fail-safe devices to prevent errors even before they can occur. Cleaning up and organizing the work areas is an excellent place to start a JIT project. Finally, suppliers must be certified to eliminate the waste that exists between buyer and supplier. There is much more that can be said about the need for maximum quality in the Just-In-Time environment, but that can be found in the numerous references in the quality section in the Bibliography.

Just-In-Time Action Plan—Quality

List those items you discovered in this chapter that should be added to your JIT action plan:

1. _____
2. _____
3. _____
4. _____
5. _____
6. _____
7. _____
8. _____
9. _____
10. _____
11. _____
12. _____
13. _____
14. _____
15. _____

Chapter Seven

Factory Flow

We finally arrive at the shop. This chapter introduces the concepts, rules, and applications of Just-In-Time "pull"systems. Typically JIT systems are most visibly seen in the shop. Other areas like—purchasing, engineering, production control, and quality—are not seen as clearly even though their actions are, or should be, "pulled" as the materials are moving through the factory. Remember, waste reduction, distance to travel, excess inventory or resources, operating expense reduction, and other issues are just as necessary in other departments as they are in production. In this chapter you will review the following concepts, rules, and applications:

1. Warehousing and traffic requirements to support Just-In-Time.
2. Establishing supplier network signals.
3. "Sight" management or signaling systems for flow simplification.
4. How to convert from work orders to daily schedules.
5. Finding the ideal JIT lot sizes to run.
6. Just-In-Time "pull" system and how it works.

Be sure to mark any area that needs improvement in your factory in the "action items" list at the end of the chapter.

> Question 19. "Transportation networks consistently deliver mixed loads from local and long distance sources."

The argument against JIT usually is tied to the cost of transportation of all those small lots on a daily or weekly basis. To consistently obtain mixed loads from local and long distance suppliers would be an expensive undertaking if it were poorly planned. The object is to find ways of consistently supplying timely small lot quantities to the actual users at the lowest total cost.

The lowest cost is generally a full truck load. The lowest cost, of course, would be regularly scheduled mixed loads on the supplier's truck unloaded by the driver through point-of-use doors directly to the users in reusable standardized containers. The cost of this can be minimized if contract carriers are used—as long as there is enough consistent volume to make it worth their while. Planned routes called "milk runs" can be defined that optimize the truck's time, and excellent prices can be obtained based on the volumes negotiated and the typical dead time when the trucks are waiting for other business. Long distance costs and deliveries can be handled by combined or mixed mode shippers. These carriers will do milk runs in a distant city, combine the loads, and ship over the road, on rail by "piggy-back," and by using any and all modes of transportation most cost effective and timely.

Often in more sophisticated JIT plants, point-of-use doors are installed to allow the drivers to unload directly to the lines without having to move the materials all the way through the plant. This facilitates faster delivery into the plant and saves the driver time. Point-of-use doors are cut into the walls all the way around the plant, and drivers are informed which doors to make deliveries to for each product. The trucks should be side loading, if possible.

The materials should be in small lots or standard containers of not over about 45 pounds. They can be easily hand unloaded without the need for material handling equipment.

Question 20. "Standardized containers with exact quantities are used between supplier and plant for over 50 percent of the volume parts."

Standardized or reusable "turnaround" containers are a key success element for many JIT plants. The least number of different type containers is desirable. They should be able to be divided internally and mixed with other containers for stacking and transport. The key concept to remember here is that the last person to touch the parts was in the supplier's production area and the next person to touch the parts will be the user of the part. All of the costly packaging and multiple handling will be eliminated. Question 22 will address the value of the containers as signals for replenishment.

> Question 21. "Weekly or daily delivery of 80 percent or more of production materials are made to the plant and directly to the production line points of use."

One of the key success elements of JIT manufacturing is the continuous flow of small lots of materials. Weekly deliveries are typical for goods from great distances. Some companies have found success with daily, twice-weekly, or every other day deliveries. There are certain economical trade-offs about having small lots delivered so often. Figure 7-1 shows the difference between large lot sizes and small lot sizes as they impact inventory costs. Notice the reduction of inventory levels when the lot sizes are reduced.

Many companies have experienced this phenomenon. As the lot sizes get smaller, the inventory also begins to take up less room on the shop floor. As soon as the inventory is removed, move the machinery and assembly operations closer to each other so parts can be passed immediately without the need for material handlers. So, as the inventory is reduced, there is more space, more efficient use of space, less material handling, and greater visibility of problems. Problem visibility is enhanced because people can see when the flow stops.

> Question 22. "Supplier deliveries are scheduled by the production processes' demand for parts."

In the shop floor, there are those who supply parts and those who consume parts, or, in other words, customer–supplier partnerships. Conceptually, each person in the supply chain is both customer and supplier. The flow continually moves through the chain based on a set of expectations. The customer expects on-time delivery and a quality product. The supplier expects to be paid for his or her effort in both money and continued business. Of course, the production line workers don't pay their "internal

FIGURE 7–1
Lot Size Impacts

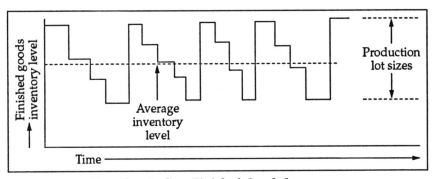

Reducing Lot Sizes Can Reduce Finished Goods by as Much as 50 Percent

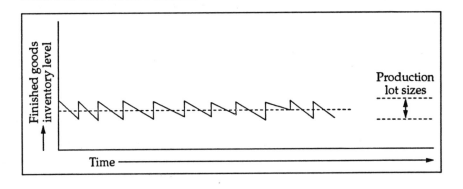

suppliers," but the company does in their behalf—and, if the "internal supplier" doesn't provide quality or delivery, they may be out of a job.

This same concept extends naturally to the "external" suppliers or vendors of parts for manufacturing. The parts are pulled from the suppliers just like they are internally. However, there are a few precautions to take before expecting the external suppliers to join in the pull system.

The supply of parts must be consistent and repeatable. This means that your company has to provide a stable schedule over a long term in order to prepare the supplier for the continuous

FIGURE 7-2
Vendor Process to Process Flow

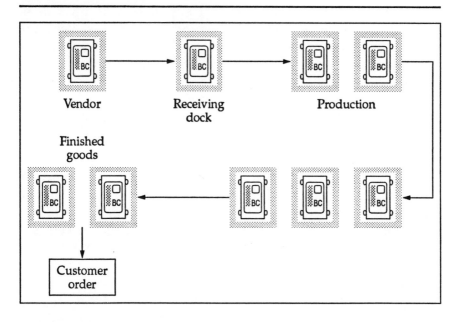

demand. Once the supplier has a clear picture of the overall demand and recognizes the need for daily or weekly shipments, then he or she can be prepared for the demand signals coming from the production system on a daily basis. Figure 7-2 shows how the external supplier fits into the internal customer-supplier chain.

Notice that the external supplier and the external customer are simply hooked at the front and the back of the flow. Suppliers will be driven by pull tickets or kanban cards for daily or weekly deliveries. When they arrive at your company, they will deliver the parts either to production or to the warehouse, where they will pick up either a production kanban or pull ticket, which is numerically sequenced. The card acts as a purchase order, supplier's sales order, receiving slip, and supplier's invoice. They return to their plant, provide the ticket data to their sales department, accounting department, materials department, and production department. Just like a blanket order, they will prepare for the next shipment.

The pull ticket that the external supplier picks up will be obtained in two ways: (1) from the production process itself and (2) from production management or scheduling who may give them an extra pull ticket to add an extra lot to the process. There are few contacts with the purchasing department.

> Question 23. "Standard containers hold exact consistent quantities and are maintained throughout the shop."

Standard quantities or lot sizes should be a fixed number for each part in the process. This means every movement of that part is in a standard container with the same amount of that part in it. This makes the parts easy to control, count, and see. Excesses of parts can be easily seen when the flow gets out of balance or someone has pushed product forward against the rules. Standard containers also should be used for suppliers so they can be brought directly to the production line without handling or repackaging.

> Question 24. "Production build and pull quantities are calculated to support only average daily demand."

All products in the schedule that are needed in a month are divided into equal daily run quantities by dividing the monthly quantity by the number of work days. This means there are no "lumps" in the flow. All lots are small or are only one piece at a time. The shift or day's run schedule is the goal for each day. The day's run quantity is a calculation of the total demand divided by approximately 20 working days in each month—creating an average daily demand that must be produced in single units run sequentially. If large lots are released along with small lots or single units, the whole system will crumble.

A wet suit manufacturer recently experienced a $10 million loss of income because of the variable sizes of its production lots. In the garment and apparel industry, if orders are not shipped within 30 days of the receipt of the order, the orders are canceled or refused because the trends or seasons change so quickly. The wet suit manufacturer had a practice of releasing orders in variable lot sizes to the shop floor. Some lots would be very large for such customers as Sears or JCPenney, and others would be small for individual surf shops or local sports equipment stores. There were hundreds of small orders for wet suits and a few very large orders all mixed together on the shop floor. Almost every worktable on the shop floor was piled high with materials.

During one year, orders for wet suits reached $18 million. When the end of the year came around, the company had shipped only $8 million worth of wet suits. The rest had not been produced within the 30-day time window and were canceled. There was finished goods overflowing in the warehouse from all the canceled orders. The company had lost $10 million in sales simply because of the lot sizes on the shop floor.

A new manufacturing manager was hired to fix the problem. He saw the piles of work in the shop and immediately ordered that all large lot sizes be cut in half (just like in Dr. Goldratt's book, *The Goal*). Within a few months the clogged shop floor began to move more freely. Large lot sizes were reduced again and, within a few months, almost all orders were getting shipped in the 30-day time window and the company realized a $10 million increase in sales. The next step was to balance and split the lot sizes into reasonably similar small lots so that all lots would flow through the shop in about the same time.

An analysis of the volume products indicated that 45 percent of the product flowing through the shop was from only about 10 percent of the products being made. A JIT one-piece-at-a-time lot size trial model was run by sending 50 wet suits through the shop with the leads and supervisors running each wet suit from workstation to workstation while others observed and balanced the cycle times of each workstation. This crude trial resulted in greater than 50 percent reduction in throughput time. The next step was to reorient a portion of the shop that contained mostly portable sewing machines of various types into a true JIT "pull" line for producing the high-volume products. This entire process took about three days to conduct. The objective was to design a production process that would produce *something for every customer every day*. The large orders that would be back-scheduled were mixed and balanced in minimum lots of five suits and mixed with all other orders each day and accumulated for shipment on a future due date. Further issues of small lot scheduling are discussed in question 40.

> Question 25. "Shop schedules, dispatching, work orders, and expediting have been eliminated and priorities are defined by shift schedules and quantities.

History and habits will be difficult to overcome when dispatching reports, work orders, and detailed scheduling are removed from the shop. But the job of scheduling becomes considerably easier, because work orders and the associated detail schedules and dispatch lists are eliminated. Instead, inventory will be considered "free" for use without the former tight controls. Work orders will disappear, thus eliminating all those labor and material tracking transactions from workstation to workstation.

Note that in many companies today the labor content of a product is becoming smaller and smaller. Thus, there is less need to be tracking labor for cost control and efficiency purposes. The focus moves to inventory. A small savings in materials cost can be substantial compared to a small savings in labor. Notice the change in percentage of labor content versus material content over the last 70 years:

1920s	1990s
Labor 80% of total cost	Labor 10% of total cost
Material 12% of total cost	Material 60% of total cost
Overhead 10% of labor	Overhead 400% of labor

A considerable amount of effort has been spent doing analysis of labor and efficiencies in the last 70 years to improve profits. When labor was at 80 percent of total cost of goods sold, a saving of 10 percent would create a major improvement in profits. A 10 percent savings in materials would be insignificant. Today, a 10 percent savings in labor would be insignificant, whereas a 10 percent saving in materials would be significant.

In the 1920s, because labor was the largest component of the cost of product, overhead expense was tied to it. The rationale was that, as labor increased, the amount of overhead would increase in relation to it. In essence, the cost of facilities, factory management, and indirect expenses could be apportioned as a percentage of labor. So, overhead was determined by multiplying the labor figure by a factor that represented the generalized or average cost of overhead.

The trends have changed. Now the labor costs are so low that a small change in labor tends to cause a large change in the cost of overhead. In the 1920s, if labor varied by 10 percent the impact on overhead was less than 1 percent. Today, a 10 percent variation in labor would mean a 4 percent change in overhead expense. The following calculations exhibit this point:

Labor and Overhead Impact from a 10% Change

1920s Labor cost = 80% times 10% equals an 8% impact on total cost.
If overhead is 10% of 80% (or 8% of total cost), then a 10% change to labor becomes a 0.8% change to overhead.

1990s Labor cost =10% times 10% equals a 1% impact on total cost.
If overhead is 400% of 10% (or 40% of total cost), then a 10% change to labor becomes a 4% change to overhead.

In this example, there is a five times greater impact on overhead than before. The smaller the labor content becomes, the greater the variation in the overhead expense when a change in direct labor occurs. In some companies, this variation could mean the difference between profit and loss on their financial statements. There has also been a major trend toward automation and the use of information systems. This has created a shift of employees and skills from lower-cost direct labor to higher-cost general and administrative positions. This shift has created disproportionate overhead costs.

So, many companies today whose labor costs are below 10 percent are considering putting the cost of direct labor in the overhead category and treating it like an expense. Then the overhead expense can be attached to the materials cost. Then overhead can be calculated as a percentage of materials cost—thus cutting down on the variation in overhead expense. Note, however, as prices vary for the materials, that a fluctuation of overhead will also occur. The current trends in cost accounting are indicating a strong need for redefinition. There are other approaches beyond the scope of this book that can be found in the area of "activity-based costing" and "process costing" in the Bibliography.

A note about company size. As companies grow, labor generally declines and indirect labor and overhead expand as more automation is used. Small companies will not experience the above costing issues as much until they have grown larger.

Question 26. "Parts are only produced as required by demand and are built in quantities approaching one."

The key to a JIT pull system is to produce to exact customer demand, no more and no less. MRP systems and older reorder point systems "pushed" products forward (or downstream) without regard for the next workstation's needs or space requirements. Because of imbalances in the speed or cycle times between workstations, there would always be someone pushing inventory forward creating inventory buildup along the flow path. Just-In-Time systems only pull forward what is needed based on four very simple rules:

1. Produce *only what has been taken* from your workstation.
2. Only produce the next work *after* the last work has been taken.
3. When your finished making what has been taken, *stop* producing.
4. *Don't push* anything forward.

This is the heartbeat of JIT flow. When the cycle times for each workstation are equalized, the whole system will be running at maximum (or, at least, of the slowest workstation).

Question 27. "Labor is not "kept busy" by building product when not needed at the next operation."

Keeping people busy is the quickest way in the world to create excess inventory. Old patterns return very quickly when people are seen "sitting around" during their idle time, waiting for the next operator to take the part that authorizes them to build another one. So, the question arises, "What should we do with these idle people?" There are a number of things to do. You could have them:

1. Help the slower workers do their production.
2. Clean up the area.
3. Do some training.
4. Do preventive maintenance on the equipment.
5. Work on continuous improvement projects.

Just-In-Time Action Plan—Factory Flow

List those items you discovered in this chapter that should be added to your JIT action plan:

1. _____
2. _____
3. _____
4. _____
5. _____
6. _____
7. _____
8. _____
9. _____
10. _____
11. _____
12. _____
13. _____
14. _____
15. _____

Chapter Eight

Production Processes

Just-In-Time cannot exist on its own in the production department. JIT requires all departments to support the production processes. A support department like engineering, which defines the products and processes, cannot be in error as it typically is. Cost accounting must develop flexibility in their data collection mechanisms. Marketing and sales must provide a more accurate and timely demand input. Quality assurance must be able to change its function from internal inspection to external auditor and trainer. Other questions address these issues. In this chapter, several points will be discussed regarding the following issues. The reader will be shown:

1. How cellular flows operate in JIT environments.
2. The impact of setup reduction on JIT throughput speed.
3. How to measure cycle time and use it to improve product flows.
4. The importance of preventive maintenance.
5. The role of manufacturing and industrial engineering in JIT.
6. The role and importance of information systems management.

Note: Keep a list of the detailed items in this chapter that are missing in your own operations and add them to the JIT Action Plan at the end of the chapter.

> Question 28. "Production lines are grouped into product family (group technology) cells or lines."

The greatest advantage that a supplier can gain over its competition is to tie customer needs directly and visibly to the produc-

FIGURE 8-1
U-Line Layouts

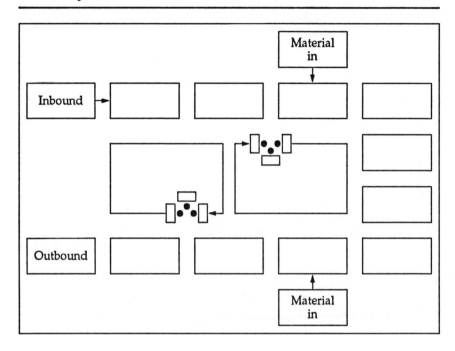

tion process, and then promise from it. This requires grouping demands into product families, which use similar processes, sometimes called "group technology." Group technology simply means identifying the "sameness" of parts or processes. It groups parts into families that will process through manufacturing "cells" of groups of equipment that process similar family parts. Essentially, by grouping all those parts that use the same type of machinery or assembly, it is much faster and easier to change schedules, setups, and raw materials to support customer needs.

Manufacturing cells may be composed of a single machine, several machines, or several operators or assemblers using the same tools or equipment. All of the equipment and operators must be in the same flow location. Figure 8–1 shows a typical manufacturing cell. Note that the operators and the equipment are organized in a "U" shape. This provides the shortest distance to move and allows the operators to help each other. This type of cell requires similar parts requiring relatively few major setups; parts need not flow to every machine. They can skip operations or backtrack within the cell based on their uniqueness.

> Question 29. "Processes support flexible mixed model runs with minimum material handling."

Techniques for mixed model scheduling are addressed in question 40. Once scheduling is done, based on family or group technology, the cell must be able to run the smallest lot possible with minimum setup or changeover (questions 33 and 34). Three issues must be addressed to achieve effective mixed model production:

1. Rapid setup. The purpose of setup reduction in Just-In-Time manufacturing is to be able to run small lot sizes. The ideal goal of setup reduction is zero time, or, at least, it should be such a minimum time that it won't hold up the flow of other parts through the cell (see questions 33 and 34).

2. Flexible cross-trained operators or flexible automated equipment. The key to good cellular manufacturing of mixed models is the ability to shift the resources instantaneously from one product to the next. Visualize this: As the next product moves into the cell, the third operator moves to position 4. Operation three will not be used for this part. The second operator rotates the parts bins on a turntable to prepare for the new product. The fourth operator moves to the first position to assist the first operator. As soon as the next product arrives, they will shift again as the part is available to be pulled to the next station. A matrix can be built to define which operator is skilled in which operation in a cell configuration. It may be of some help in restructuring the cells each time a new product hits the cell.

3. Timely minimum lot inventory that is easily accessible. At first, there will be various amounts of inventory on the floor between operations, sometimes called "kanban" inventory. As setups and cycle times are balanced and reduced, the inventories can be removed. As soon as production smooths out, remove some of the inventory from the kanbans between stations.

This process was done recently in Safetran Systems, a JIT startup company, with dramatic results. One of the processes—an assembly line or cell that made a single product—had 130 modules or printed circuit boards, in various stages of assembly scattered along the assembly lines on rolling carts. The leadperson determined that the modules being worked on at each workstation were enough to keep the flow going without using the kanban racks. So, he presented the idea to the group in the weekly JIT problem-solving meeting. He proposed to roll the kanban racks, which were located between each workstation out of the line. Everyone was to work only on the items they had and to follow the pull system rules (never push anything forward, don't work on the next item until yours has been taken). A cycle time analysis was done for each station by the workers, and it was determined that the line would function smoothly with only a small amount of idle time in a few stations. The team agreed to try the new plan.

The experiment worked. The inventory was removed with essentially no major problems seen on the line. The team was very nervous at first, but quickly adjusted to the new idea. The inventory was reduced from 130 units to 20 units, or an 80 percent reduction of inventory—with no negative impact on production. The net dollars reduced was $14,300. This same process was done throughout the plant, resulting in many similar reductions in work-in-process inventories.

>Question 30. "Manufacturing is actively involved in product and process design improvement for quality and producability.

The people most knowledgeable about how the product is actually made are the frontline production workers and their immediate leaders. The manufacturing or process engineers also have great skills and experience with process design and problems. But the frontline workers must be included in the frontend definition of a new or redesigned product. The following is a model of a producibility or manufacturability checklist that might be employed to incorporate the production experience into the design stage. Implicit in this activity is the fact that both the product design and

FIGURE 8-2
New Products Introduction Cycle

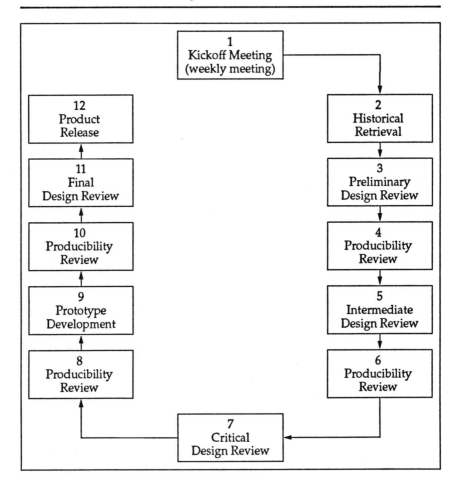

the process definition must be done at the same time. Figure 8-2 shows a cycle for introduction of a new product into production and the associated producibility checks that would occur during the cycle.

The sample list in Figure 8-3 applies to each of the producibility check points in 8-2.

FIGURE 8-3
Producibility Check List

Item to Check	Producibility Check No.			
	1	2	3	4
Item master data load	X	X	X	X
BOM data load	X	X	X	X
Drawing release	X	X	X	X
Sketch release	X	X	X	
Quality specification release	X	X	X	X
Tooling/Fixture design schedules	X	X	X	X
Purchase of special dies, molds, etc.	X	X	X	X
Standards definition	X	X	X	
Test—Manufacturing samples	X	X		
Process/Routing operations defined	X	X	X	
Production planning data projected	X			X
Test equipment design	X	X	X	X
Material handling equipment specs	X	X	X	X
Special materials handling requirements		X	X	X
Drawing/Model/Part number control	X	X	X	X
Methods, handbook/Instruction manuals			X	X
Facilities planning	X	X	X	X
Plant layout/Changes	X			X
Manufacturing approach (MRP, JIT, etc.)	X			
Common part/Component options defined	X	X	X	X
Make or buy decision	X	X	X	
Lead time requirements	X		X	X
Service parts requirements	X		X	X
Catalog, price lists, schematics	X	X		
Preproduction change control	X	X	X	X
Parts/Component list	X	X	X	X
First article inspection/Acceptance			X	
Production release				X

At least two other mechanisms can be used to design and release a quality part that is easily and economically producible: design of experiments and failure mode and effects analysis.

These two items can be found in the references in the Bibliography at the end of the book.

Question 31. "Production parts have been designed to facilitate fast changeover."

Changeover or setup is the process of changing from one product to another. In the Just-In-Time environment, the ultimate objective is to produce one unit at a time of each product the company sells every day. To approach this ideal, the objective of zero setup time becomes the driving factor behind both setup reduction and product design. Product design essentially defines process design, also. Process includes the equipment, tooling, employees, or material handling requirements needed to build the product. Often the process definition is not considered in the product design. Yet, the process itself is where some of the greatest waste occurs, simply because of lack of foresight.

Many techniques can be used to facilitate rapid changeover, such as mistake-proofing or "fail-safing (the Japanese call it poka-yoke, see question 15). The use of simple design features like guide pins in the part or the fixtures to ensure correct orientation are common. There are many examples of parts being designed with interfacing parts oriented in a manner that they can never be reversed or inserted wrong. In many cases, just changing the type of screws or fasteners would facilitate faster changeover. Small notches can be used to guide parts for accurate locating. Recently in a plant a cumbersome transformer bracket was redesigned to be in two sections instead of one to facilitate faster changeover and assembly. It reduced the changeover time and associated assembly time by 60 percent. Concern must be to continually review setup and changeover reduction ideas from a design point of view.

Question 32. "Tooling and fixtures are available when setups and jobs begin."

In most companies, it is easy to reduce setup time by 50 percent or more without ever making any changes to the tooling or the product. A majority of the time spent on many setups is spent in travel time, preparation of the tooling, cleanup, finding parts, tracking down misplaced fixtures, and a myriad of other time wasters. A well-known doorknob manufacturer recently went through a setup reduction program in an attempt to organize its setup process. Setups were taking an average of 16 hours apiece. The approach taken was simple. They made sure everything was ready when the setup began.

A setup–teardown card was designed that had all the teardown data, instructions, measurements, and tolerances on one side of the card. The setup data, instructions, measurements, and tolerances were on the other side of the card. Every time a setup was torn down, readings from both the tooling and machine as well as observations and comments were recorded on the card. The card and all of the parts were returned in a kit box to the local setup room, where the tool was inspected, repaired, refurbished, prepared for the next setup, and placed in a schedule position on a shelf designed to match the master schedule. When the setup was initiated, the readings, tolerances, and measurements were recorded on the other side of the setup card for reference on the next setup. All of the parts, shims, gauges, and tools were included in the setup kit box so nothing would get lost. The kit box was checked and associated tools were issued by the tool room kit-prep process. The results of this process alone reduced the setup time by over 50 percent to an average of about eight hours per setup. Unfortunately, as business slowed, the company laid off the tooling room person who prepared and inspected all the tooling and the system fell back to an average of 16 hours per setup. How easy it is to fall back to the old ways.

Question 33. "Rapid setups are established (less than 10 minutes) for most machines and lines."

Setups are a nonvalue-added activity. Setups are costly, time consuming, and wasteful. They do not add anything of value to the product for the customer. They should be eliminated. However, it doesn't work that way, so the best we can do is to reduce them to the least time possible.

Why do we want to reduce setups? *Not to reduce the number of setup people, or to produce more product, but to be able to run smaller lot sizes and to improve the quality of the setup.*

One of the most common approaches for setup reduction is to

do as much of the setup as possible while the machine is running. This alone could have a big impact on setup reduction. There are two elements of a setup: internal and external. *Internal* means any activity that can only be done with the machine shut down. *External* means any element that can be done while the machine is still running. Simply list all steps that are currently being done during shutdown. Identify all elements that *could be done* while the machine is running, change as many internals to externals possible, and retrain the setup people.

Note: there are a few rules about setup reduction to remember. The measurement for setup time is the time from the last good part that ran on the last setup until the first good production part run on the new setup. Obviously, the time measure isn't taken if the machine is idle. Don't spend much time doing setup reduction analysis and changes if the operation or machine only runs occasionally. If a manufacturing cell typically runs the same parts with minor changes, there may not be any setup at all. It may be that a setup is completely unnecessary. If it is necessary, then do an internal to external analysis.

One of the tricks people have done is to use a video camera to record the setup activities and then to replay the tape in a workshop session focused at reducing the setup. This is an excellent way to review some of the typical concerns that effect the setup, like the machine itself, the fixtures or attachments being used, the tooling, the material to be processed, the surrounding work area, and the work routines. Each of these must be considered in a setup reduction program and can be easily recorded and played back for analysis and training.

A check of the standard routings must also be made to insure that the correct processes are being used in the costing system. The regular routines and patterns for setups, methods for adjusts, and checks and practiced routines with documented methods must all be investigated, simplified, and standardized.

The ideal setup time is zero time, zero cost, or even a completely automated setup. However, a setup that takes less than 10 minutes is usually a reasonable time. Shigeo Shingo describes setups as "single minute exchange of die" (SMED) in his book, *A Revolution in Manufacturing*.

The following is an analysis of the setup impact on the cost and lot size or run quantity of a part:

Situation 1—Lot Size of 100, No Setup Reduction

Lot size	= 100 parts
Setup time	= 120 minutes
Process time per part	= 5 minutes
Total time per part	*= 120 min + 100 parts × 5 min/part ÷ 100 parts*
	= 6.2 minutes per part (620 ÷ 100)

Situation 2—Lot Size of 50, No Setup Reduction

Lot size	= 50 parts
Setup time	= 120 minutes
Process time per part	= 5 minutes
Total time per part	*= 120 min + 50 parts × 5 min/part ÷ 50 parts*
	= 7.4 minutes per part (370 ÷ 50)

Situation 3—Lot Size of 50, with 50% Setup Reduction

Lot size	= 50 parts
Setup time	= 60 minutes
Process time per part	= 5 minutes
Total time per part	*= 60 min + 50 parts × 5 min/part ÷ 50 parts*
	= 6.2 minutes per part (310 ÷ 50)

In these examples it is clear to see the impact of setup reduction on the ability to run smaller lots at the same cost as the larger lot. This adds flexibility to meet variable customer demand. It also reduces inventory by 50 percent of the parts in process.

After cleaning up and organizing the basic setup and teardown procedures, then more mechanical or automated changes of setup methodologies should be performed. Some examples might be:

- Roller tables for dies or tooling that are too heavy to lift are designed to match the height of the machine.
- Installing rollers on the table and machine so the die can be pushed into place without the need of two people or material handling equipment.
- Installing locator pins in the machine that stop and center the tooling when inserted.
- Replace threaded bolts and clamps with air pressure clamps or swing bolts to speed up fastening and eliminating

dropped or lost nuts and bolts.
- Premount tools in spare tool holders.

There are hundreds of examples in the *Poka-Yoke* book listed in the Bibliography.

> Question 34. "Cycle times of each workstation, cell, or line are matched to upstream and downstream times."

This is the ultimate goal of Just-In-Time continuous improvement: to obtain a balanced line of equal cycle times throughout the entire manufacturing flow of a product. If each of 12 operations took two minutes to perform its job, that would mean one part would be completed every two minutes. However, if *only one* of those operations took three minutes, the entire process could produce one part every three minutes. That's a 50 percent increase in cycle time. Cycle time is the time between the completion of two consecutive parts from a process.

It is important to understand that the rate of output is the rate of the slowest operation. If that one slow worker can only produce one every three minutes, the whole line must wait for him or her. If, however, two people could split that job, the time could drop to one and a half minutes per piece and the rate of output would drop back to two minutes between parts. Thus, if it is desirable to increase throughput from a line, the constraint or slowest operation must be found and every effort applied to reduce it to the next slowest rate. Then, the next slowest rate becomes the focus of reduction efforts. This process should be unending.

Any product line that is matched with its upstream and downstream lines would be somewhat temporary and would be continuing to improve. If the cycle times are matched, that means that work is in process to improve throughput.

Cycle time analysis is usually accomplished by building a flow chart of each of the steps in a flow and recording times for each of the steps similar to the example in Figure 8–4.

It is very easy to detect the time constraint in the system in Figure 8–4. It's in the inspection area. It takes 16 hours just waiting for inspection, unless there is some kind of failure with the part, then it might take 32 hours to get it cleared or rejected.

FIGURE 8–4
Flow Charting Sample

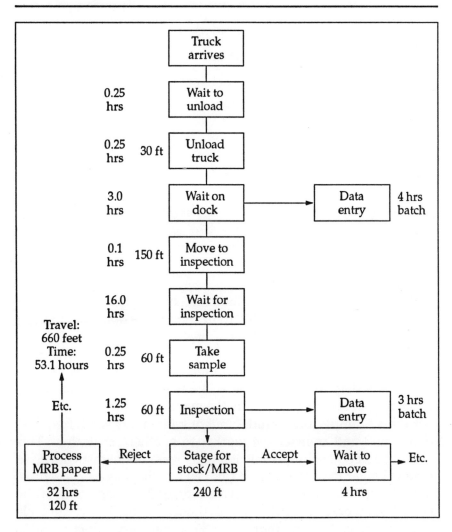

Consider the amount of overall cycle time, called "tact" time, that this flow has if the part is a good part—about 25 hours. Imagine if the wait for inspection and the wait to move to stock were eliminated. The tact time would drop by 20 hours. If the rest

FIGURE 8–5
Intercell and Intracell Balancing

Any unlike operation will imbalance the cycle times inside or between cells.
Work Station 1   Run Cycle Time   Required Cycle Time
60 seconds
Work Station 2
45 seconds   Idle Time   15 seconds
Work Station 3
Idle Time
15 seconds   45 seconds

of the steps were reduced or eliminated by certifying the supplier and delivering the parts directly to the production line, the tact time would drop to only a few minutes.

This technique will help to highlight the cycle time differences and allow decisions to be made to rebalance the workers. Figure 8–5 shows that some idle time will always occur in cells if any imbalance of speeds exist inside the overall tact time.

In this example, one solution might be to move work from station 3 to 2 and transfer the worker to another constraint. This would be a good solution only if an upstream or downstream cycle time is longer than this cell. If this cell has the longest cycle time, then the solution might be to have worker 3 offload some of the work from worker 1 for 15 minutes, thus reducing the overall cycle time for the cell to one part every 30 seconds—a reduction of 33 percent.

Cycle time analysis can be applied to any operation or office process inside the company or at a supplier. This is an extremely valuable tool. Remember, the entire rate of output from any system is at the rate of the bottleneck or slowest operation. Thus, any improvement in the bottleneck is a direct improvement in throughput. How will you know what your bottleneck is if you don't do a flow chart?

> Question 35. "Manufacturing processes have been reoriented to eliminate material handling."

The natural consequence of tight, low inventory pull systems is the elimination of inventory from the shop floor. Many companies have reported 80 percent or more reductions of inventories. If this were the case, there has to be an equivalent increase in available space around and between operations in the shop.

By cleaning up the shop and so organizing it that only the necessary tools and equipment are in their proper place, and eliminating the storage of most personal paraphernalia owned by the workers, much more room becomes available. Once this extra space is freed, it provides an opportunity to bring the workstations closer together. As they come closer together, the amount of material handling equipment—like pallets, shelving, storage bins, boxes, forklifts, aisleways for forklifts, and other space-consumers—is reduced. These items are all waste and should be eliminated.

With the elimination of space wasters, the production flows can be reconfigured to much less space, and, this essentially facilitates the capability of making only one unit at a time.

> Question 36. "Process problems are identified and visibly signaled immediately on discovery."

The heart of Just-In-Time manufacturing is its signaling system. It operates everywhere. Visibility is critical because everything is essentially synchronized, and, if something stops moving, the whole system must see it immediately. Immediate feedback is a key feature in the signaling systems. Stopping the line is a common signal seen, usually by a light or horn. However, the most common signal by far is the signal to move or produce a part in process. These, of course, are the kanban cards, squares, containers, and so on.

The example of a manufacturing cell in Figure 8-6 illustrates the impact of signals on the flow through a cell. One of the rules of Just-In-Time is that one cannot produce until the part just completed has been pulled to the next operation. One of the deficien-

FIGURE 8-6
Pull System Stop Signals

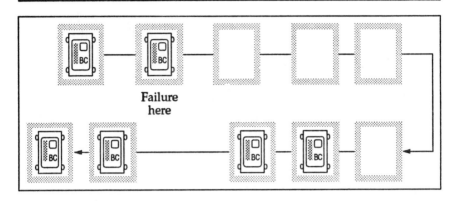

cies of this rule is that, if a failure occurs upstream and the downstream operators don't know about it, the entire line will be stripped clean downstream of the failure. If a stop signal were used, the whole line would stop when the problem occurred, thereby creating immediate available help to offload the work from the failure. If the manufacturing and quality people are signaled, they can be immediately available to find a solution.

If the line continued to run, the line would be empty and the workers would be idle. On the other hand, if the line stopped, there would still be idle workers, but they would be able to support the troubled workstation more readily than if they were working on product that shouldn't be made yet.

As long as the philosophy of continuous improvement is in place, the line stoppages will continue to reduce in both severity and time. One of the General Motors plants in Flint, Michigan, tried a different approach to shutting down the line. The plant found that many of the workers were stopping the line for minor problems that could have been repaired without stopping the line. So, the plant installed a second *yellow* cord that could be pulled so workers could call for help without shutting down the line because of minor glitches.

Question 37. "Manufacturing engineering is located in the production area and is immediately available for problem resolution."

Manufacturing engineering is responsible for defining, building, and monitoring the proper processes needed to build a high-quality, cost effective product on time to meet customer needs. So, it doesn't seem to make a lot of sense that these workers would be located in some remote engineering department unable to be reached when process problems occur. The engineers need to be as close to the processes as possible so they can respond quickly to quality, material, or process problems created by the design of the product or process. In the JIT process where continuous improvement is always present they must be available at all times.

Question 38. "Scheduled preventive maintenance is considered an important part of production performance."

There is no good reason for down time on a Just-In-Time line caused by poor maintenance. Yet, this area seems to be last to be implemented in many companies. Often, the cost for downtime is thousands of dollars per hour as well as late customer deliveries, inventory buildup, expediting costs, unplanned changeovers, and wasted labor.

Good preventive maintenance is the key to the reduction of emergency downtime. Emergency downtime is when any process stops and is unplanned. Downtime should be tracked like setup time: from the time the process went down until the time it is running good parts—*from the last good part to the next good part*.

The responsibility for good preventive maintenance is spread throughout the manufacturing organization, not just the maintenance department. The operators must be trained to do daily maintenance and checks as part of their jobs. The maintenance department has the responsibility to lay out and define all the necessary maintenance activities for all the machines in the plant and then promulgating preventive procedures and training all the operators.

Preventive maintenance is any action that will prevent machinery breakdown. It ensures the quality and reliability of the process. It can be organized in a number of ways. Apart from training operators, many actions require repair, replacement, or analysis. These activities could be scheduled on an offshift, on weekends, early morning, or late shifts so there would be no impact on production flow.

However, many times there are overhauls, line changeovers, or routine replacements that take longer than one or two days. These activities must be scheduled directly with the master scheduling group. By releasing a mock demand item equivalent to the typical time that a volume of product would run in production, the schedule can be maintained and balanced by the scheduler without interrupting the planned flow.

Using this technique, the schedule will be maintained on the regular products. However, there must be advance notice. Another way to plan maintenance downtime is to have the scheduling system always schedule a specific amount of time each week for each of the lines needing preventive maintenance. There is usually resistance to these techniques because they tend to "limit" production. Remind the objectors how much it costs for emergency downtime.

A paint manufacturing facility recently reduced emergency downtime by 52 percent merely by doing the preventive maintenance offshift and building an accurate inventory of spare parts and supplies to support the planned maintenance.

If SPCs are absolutely necessary for JIT, then preventive maintenance is an absolute necessity to support SPC. As the processes are refined, lots reduced, setups reduced, and buffers eliminated, then nothing is left to protect the flow. This recognition will help supervision to get the operators to use their expertise and experience to pay more attention and to conduct preventive maintenance checks (PMs) on their own machinery, rather than waiting for maintenance people to do it. This develops ownership if the operator understands that the buffers are gone and there is no protection. Their knowledge of the machinery is extremely valuable. The bottom line is still that, to build high quality parts the people and the machinery must be operating at peak quality performance.

Total productive maintenance (TPM) is an extension of these concepts and should be investigated by the reader. Check the Bibliography.

Chapter 8/Production Processes   99

Just-In-Time Action Plan—Production Processes

List those items you discovered in this chapter that should be added to your JIT action plan:

1. _____
2. _____
3. _____
4. _____
5. _____
6. _____
7. _____
8. _____
9. _____
10. _____
11. _____
12. _____
13. _____
14. _____
15. _____

Chapter Nine

Master Planning

The Just-In-Time Self Test book highlights only six elements of the master planning process. These six questions are probably the most important in the Self Test because, without accomplishing the master planning functions, there will be very limited real performance from the rest of the JIT implementation activities. The Self Test is a test of the physical performance of your company. So, the output from production measured against a plan is the key ingredient to master. Obviously, if there is no plan, there is no measurement.

>Question 39. "Daily rate and level schedules are used and meet due dates."

Daily rates and level schedules are almost impossible to meet without a formal production plan (or sales and operations plan). This is a separate high-level integrated planning document managed by the key top managers in your company. It coordinates and balances three major conflicting customer demand elements into one cohesive document. Those elements are: sales forecast, customer backlog, and master production schedule (MPS). It's very easy for each of the elements in this demand planning chain to get out of sync with each other. Figure 9–1 shows an overview of the planning flow and the interfaces for master planning. *Note:* product families in the production plan are the origin of all subsequent lower-level scheduling. The next several pages will address each of the elements in this overall master planning flow chart.

Driven by a business plan derived from a long-range forecast, the production plan should typically project for a year or more into the future. This plan is composed of product families. Each of the family groups should be grouped by the capacity that will be used

FIGURE 9-1
Master Planning Interfaces

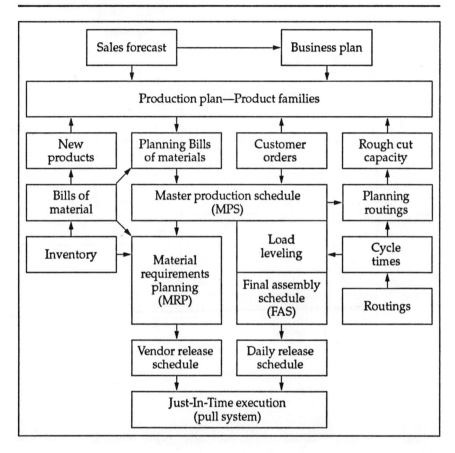

to produce them. Generally, a "planning router" is used to identify and define the similar stages or work centers (cells) of sequential production capacity that those products go through. In the job shop environment, the family groups may not be as clearly visible. Even if the families are not definable, look for product groupings or major assemblies that use the same capacity, or similar routers, as a starting point.

The production plan, normally expressed in monthly increments, is broken into weekly and daily increments through the

FIGURE 9-2
Production Plan and Master Schedule Integrity

	Scheduled Products	MSP by Weeks 1	2	3	4	MSP Total
Product group A	A-1	10	10	10	30	60
	A-2	40	20	20	10	90
	A-3	0	30	30	10	70
	MPS group subtotal	(50)	60	60	50	220
	Production plan quantities	60	60	60	50	230
	Cum variance to plan	–10	–10	–10	–10	(–10)
Product group B	B-1	10	30	10	20	70
	B-2	30	10	20	40	100
	MSP group subtotal	40	40	(30)	60	170
	Production plan quantities	40	40	50	50	180
	Cum variance to plan	0	0	–20	–10	(–10)
	MPS performance to plan					(–20)

master production schedule and the final assembly schedule (FAS).

The MPS should be the focal point in any company implementing JIT, because it plans and controls the time sequences of the overall resource utilization throughout the company. Once the production plan is in place, basic mathematical calculations can be used to compute all subsequent lower-level resources through the MPS. Top management must manage from this level to plan or replan the factory based on a realistic level of capacity. Thus, the relationship between the production plan and the master production schedule must be cemented together. Here's how: in Figure 9-2 the total demand of all components in week 1 for family "A" is 60 units. This *aggregate* total of all products in the "A" family must be matched by the master production schedule. Although the individual scheduled quantities may vary, the aggregate total must match the production plan family group quantities.

Meeting this number means that the plant produced the total family requirements as planned by utilizing the capacity as planned—essentially trying your plant's specific capacity directly to the customer's needs. However, there may be some capacity problems later on unless a preliminary capacity check is done at this level. A "rough cut" capacity plan can be generated from the master production schedule, which can be used to validate the realism of the production plan. Good references are available on these techniques in the Bibliography, such as *Manufacturing Planning and Control Systems*.

When there are a high number of master scheduled items, a "planning bill" of material can be used to simplify the forecast/master schedule interface. A master schedule can be calculated from a family group in the production plan by structuring a bill of material whose "end item" is a family group, and the "components" are the actual master scheduled items. In this case, the quantity per "end" item is the forecasted percent of the total demand that each "component" represents. These components are the actual calculated quantities of the "real" items that will be scheduled in the master production schedule. Thus, for Family Group A, items A-1, A-2, and A-3—whose sales represent 20, 30, and 50 percent of the total family group "A" sales—can be projected into a MPS planned requirement. These percentages and calculations are only changed when the forecast changes. Changes are not allowed in the current (week or so) period unless lead times are extremely short and suppliers are highly responsive to changes.

Let's return to the second half of question 39: " . . . *level schedules* met on due dates." Assuming that the master schedule has been stabilized, level loading of the plant on a daily basis is derived from three tools:

1. Load leveling.
2. Mixed model scheduling.

3. Cycle time management.

Load leveling is a process of planning the master schedule's most current period—called the final assembly schedule (FAS)—to build the same product mix consistently, usually in weekly or smaller time buckets. The final assembly schedule is merely the current one to five days of the MPS. This is an absolutely firm and unchangeable segment of the MPS, in which *only* real customer orders are scheduled. Its purpose is to match the customer orders to the master schedule. The master scheduler must constantly monitor and manipulate the customer order backlog to keep the orders and the schedules in balance. The FAS is usually found in make-to-stock or assemble-to-order companies.

Essentially, all products in the schedule that are needed in a typical month are divided into equal daily run quantities balanced against daily run capacities. This process makes certain that the capacity to build the schedule is reserved in advance. The monthly schedule is broken into daily rates by dividing the monthly requirements by the number of work days.

Mixed model scheduling generates the smoothest possible schedules for all internal operations, which subsequently establish the baseline for external supply systems. It sets the proper time-balanced homogeneous sequences that will be repeated consistently throughout the FAS schedule periods. See *Just-In-Time for America* in the Bibliography.

Cycle time, which is the time from one good part to the next good part, is an important part of the analysis. Cycle time management identifies and controls the time intervals between each part flowing through the manufacturing process. The longest cycle time in any flow will become the determining element for daily output from any process. No matter how fast the other operations are, the process can only produce at the rate of the slowest part.

An important focus for manufacturing and for industrial engineering and the operators is to find ways to match the cycle time in each step in the production process to each other step. Throughput can be increased by reducing cycle times of the slowest operation throughout each step within a work cell. Once this is accomplished, then the overall production cycles can be scheduled more easily and consistently.

Establishing uniform or stable scheduling is now attainable. Daily rates and level schedules met on due dates usually occur after many of the bottlenecks have begun to be repaired. This includes parts shortages, inventory errors, poor quality, inconsistent demand, and design changes as well as poor quality workmanship, new trainees, and breakdowns.

> Question 40. "MRP is used for demand planning, customer committing, supplier schedules, and supplier capacity planning."

Material requirements planning (MRP) is a calculation tool for planning mostly external resources needed for JIT performance. All priorities (ie, due dates) and quantities that are calculated by MRP are driven by the master production schedule. MRP's greatest strength is integrating the MPS to all of the lower-level requirements through the bill of materials on a time-phased basis in support of purchasing and long-term planning, but *not* the shop floor. By using a bill of material, purchase order status, inventory data and lead times, MRP calculates all material needed to support the JIT demand-driven execution at the suppliers as long as the data are very accurate. However, some changes to the typical MRP process must be made to be able to support JIT. First, let's look at the basic MRP logic. The four basic data elements that MRP uses for calculation are:

1. Requirements—typically the master schedule.
2. Open orders—typically purchase orders in a JIT system.
3. Onhand inventory.
4. Lead time.

The data accuracy for each of these data element is critical to successful MRP calculations in support of a JIT system. Errors in any of these four could create a line stoppage. The output from MRP is used to manipulate purchase priorities—either to release new orders or to reschedule them prior to receipt at your plant or into schedules at your supplier. MRP recommendations are typically sent to the buyers in purchasing, who will evaluate and act on the MRP recommendation and work with suppliers to integrate

FIGURE 9-3
MRP Logic Conversion to JIT

Lead time = 1 day Final Assembly schedule = 12/day	Period							
	1	2	3	4	5	6	7	8
Part A-1			12	12	12	12	12	12
Component X (needs 2 per)			24	24	24	24	24	24
Component X (needs 2 per)	48	48	48	48	48	48		

MRP calculates work order for 48 parts of Y to make 24 parts of X here. Then recommends a work order to make 12 parts of A-1.

Stable JIT–Type Flow Bills of material changed to two levels here and one level here. JIT takes over.

the new demand changes into their longer-range schedules.

A check of open purchase orders is made by MRP to see if more of the required item are already on order with suppliers. Then onhand inventories are checked to see if the required items might be in stock. Then, a netting calculation is done. There are many volumes written about the MRP process. This discussion is merely to acquaint you with its simple logic and to introduce you to them and the use of MRP as a support tool for JIT. Several good books are listed in the Bibliography that thoroughly cover all the capabilities of MRP.

MRP facilitates customer committing by (1) planning long-range purchased parts replenishments to support JIT and (2) reporting exceptions or changes to schedules.

MRP logic, calculations and data management capabilities are integral to a full Just-In-Time system. MRP tools are used to do the production plan, the planning bill explosions, and the master production scheduling calculations as previously discussed in this chapter. It is not necessary to plan the internal factory requirements—that's done by the JIT execution process.

If a company is converting from MRP to JIT, MRP can help calculate the resources for final assembly output so all required materials are in place. Once the process is stabilized, it can be driven by JIT pull, rather than by the MRP system. So something must be done to identify the process of conversion from an MRP system to a JIT system using the MRP tools. Figure 9-3 shows how the flow of parts in the MRP system calculation ends up looking like a consistent, repetitive JIT flow when an MRP-generated schedule is running in a stabilized condition.

The MRP chart shows the schedule for three parts: one end item and two components in a three-level bill of material. The daily quantities for the three levels of parts in the center section (highlighted) is exactly the same after the first end item has been completed (in day 3) and continuous parts flow equally.

In the beginning (period 1), the bill of material is structured in three levels. In the end (period 6), the bill of material is structured in a single level. Once the JIT execution process takes over in the center section, the bills can be "flattened" (or phantom bills can be used) to move all the MRP calculated events in the second and third levels up to the first level. This process removes MRP from the internal planning and the subsequent issuing, tracking, and reporting of work orders. MRP is then used only for the replenishment of purchased parts and the previously mentioned master planning processes. See *Just-In-Time for America* in the Bibliography.

In a JIT environment, the FAS and MPS requirements are the "work quantity" requirements grouped in daily or sometimes hourly buckets. Work orders are eliminated because there are no lower levels that MRP would drive. Then the JIT pull system manages the rest of the movement and shop flow activities. However, some manufacturers may use MRP to schedule other non-JIT processes.

The most needed MRP function in a JIT system is the longer-range scheduling of materials from suppliers. Supplier scheduling can be accomplished from the MPS and MRP. Long-range purchase orders for the stable section of the MPS can be released to the suppliers in balanced quantities. Option agreements can be utilized. Obviously, the MPS and FAS must be stable and firm during the supplier's lead time in order for them to plan and support your needs. Unfortunately, the typical lead time for a supplier's production and delivery cycle could be many weeks or months long. It would be virtually impossible to hold the MPS and FAS stable for that long. To overcome this lead time problem, many JIT companies *contract* for the supplier's capacity.

Purchase contracts can be made with suppliers, which can guarantee the commitment of their facilities to build products without specifically defining the products. This process will facilitate the release of orders at the last possible moment with the shortest lead time. Most suppliers usually have a large queue of orders in their schedules typically received on a first-come-first-served basis. By allowing a contract for a block of their production capacity, the supplier can obtain the necessary raw materials and schedule their capacity to run specific products or group of products with the understanding that anything not ordered from them will be billed at or near their cost. If you don't use the capacity as agreed, they may then choose to use the reserved capacity to build other customer's parts as long as the next block of capacity is not affected.

Question 41. "Management participates in the planning and replanning process and commits to a realistic capacity level."

The production plan, or the sales and operations plan, is the tool to be used by the top management of the company to plan, in aggregate, the overall production levels and to measure the performance of the plant. Five or six top managers must meet on at least a monthly basis to "buy off" on the projected plan and to evaluate the performance of the MPS to the current period plan. The inputs to this meeting are:

- The current production plan.
- The performance of the plant MPS to the production plan.
- The status of critical materials that would effect the plan.
- Financial data and performance to budget.
- Major marketing or sales changes since last meeting.
- New product changes that will have a major affect on capacity.
- Major manufacturing capacity changes since the last

meeting.

Remember, this meeting is designed to match aggregate performance in product groupings as they relate to aggregate capacity. This is not a detailed master scheduling meeting dealing with customer orders or internal capacity problems unless they are *global* in impact. The results of this meeting are either an agreement to stick to the plan or to adjustments to the production plan, which will take place beyond the current lead times used to plan the operations in the MPS.

Question 42. "Marketing promotes the demonstrated benefits of JIT to customers."

Marketing and sales departments have everything to gain and little to lose from a Just-In-Time implementation. They have several responsibilities that are essential for JIT success:

- Providing precise customer feedback.
- Selling the production plan as defined by top management.
- Integrating the customers into the JIT forecasting process.
- Coordinating customer visits, and building strong relationships between customer resources and internal planning processes.

There are many advantages to the Sales and Marketing groups that will help them increase sales:

1. JIT will maintain or lower costs to the customer as the lead times, inventories, and data integrity errors are lowered.
2. The focus on quality as a company policy will be seen by customers.
3. Faster deliveries can be promised and upheld as lead times decrease.
4. Flexibility to adapt to customer problems results from shorter lead times.

These advantages far outweigh the resistance that will occur as pressure for selling only to the production plan occurs. Habits are hard to break. If they have been accustomed to excess finished goods and the ability to get schedules changed whenever they ask, they will resist both having to sell only to the production plan and to provide an accurate, timely, and more thoroughly analyzed forecast.

Question 43 "Production rates exactly equal demand rates."

These questions are the proof of the master planning process defined in questions 39, 40, and 41. The master scheduler is responsible for maintaining this delicate balance. This job is not a part-time job filled by an unskilled planner. This is the essence of control by a skilled manipulator. This person must:

- Have good analytical skills.
- Be a good negotiator.
- Have thorough product knowledge.
- Exhibit common sense, and logical approaches to problem solving.
- Be good at working with people.
- Be well trained in master scheduling tools and techniques.
- Have integrity and the guts to stick to policy under pressure.

It should be noted that, for small companies, especially job shops, the production quantities—not rates—equal demand rates. This is common to job shops anyway. Typically, job shops are scheduling and running only to the orders that they have on the books.

In the repetitive environment, production rates are daily rates derived from the monthly rate and tied to the MPS driven by the production plan. It is the master scheduler's job to match these rates to the demand rates. In the JIT process, the order demand will create the pull either through a "two-bin" type of finished goods signal or by an actual customer order. A two-bin system is simply a kanban quantity that is a minimum lot size stored in two bins. When one is pulled, it signals an immediate replenishment from the shop. The second bin will cover the lead time only for the replenishment. The kanban lot sizes are determined by finding the typical minimum demand lot size for the customer base orders daily or periodically.

For example, in a dental products division of a well-known

company, the typical lot sizes run in production were about 2,500 or more brackets at a time. The typical customer order for any one week was approximately 200. By reducing the lot size to 250 they were able to supply 10 times the number of different products in the same time it took them to make one product. They were always behind trying to run the right products in the larger lots. By doing this process and by focusing on setups and other internal constraints, they were able to reduce inventory by 45 percent and increase customer service and lead time for delivery substantially.

Question 44. "Customer on-time delivery rate is 98-plus percent as committed."

The true measure of any company is its ability to deliver a quality product on time at a good price. This can be done with a good MPS that matches the production rates with the demand rates and is validated by available capacity. If this is true, then it is automatic that the delivery rate to customers is 100 percent as committed. A "no" answer to this question means you must pursue the solutions defined in the rest of the Just-In-Time Self Test.

Be careful of the word *committed*. If a promise is made with an extra lead time "cushion," you can suffer an eventual loss of business even if you can deliver as committed, because your competition will be able to deliver in less time. Often companies will commit without checking that the MPS will be balanced against available capacity. But, by expediting excessive overtime and disrupting schedules, they are somehow able to deliver products 100 percent as committed. The other questions in the Just-In-Time Self Test will help you focus on balancing and integrating all operational activities, not just suboptimizing a few at the expense of the others.

The best measure for on-time delivery for a make-to-order company should be based on the time the item was placed on the truck, not the time it went into shipping or finished goods.

Only a few very important elements must be in place for a successful Just-In-Time implementation, and one of those elements is master planning. It starts with a production plan based on product families whose aggregate demand for capacity is managed and balanced by top management. The master production schedule then segments the production plan into weekly and then daily buckets of equal demand. Through the load leveling process, the master scheduler divides the monthly schedules into equal daily or weekly schedules and then inserts specific customer orders into a final assembly schedule, which represents only a few days or weeks of absolutely stable production levels. Then, mixed model schedules are calculated to balance hourly quantities for each day in the FAS. Cycle time analysis identifies the internal unit production processing time per operation and is used as the basis for load leveling of the FAS.

Material requirements planning takes its input from the MPS and internal inventory and order data. It then recommends replenishment orders for parts. Longer-range supplier schedules are produced from MRP, which can be used to contract for supplier capacity without the commitment of actual quantities until the last possible moment. This removes the majority of supplier lead time from the overall planning cycle.

Properly designed and managed output from master planning will result in production rates that match customer demand rates and result in 100 percent delivery on time to customers.

Master planning is one of the least developed of the organizational activities in most manufacturers, probably because it is dealing with the abstract, unknown, and unpredictable nature of future customers needs. However, this is also the greatest area of growth. By continuously repeating already known status quo-oriented activities, a company can easily fall behind its competitors. Just-In-Time systems reveal the unknown and can be the cause of continuous improvement that will outdistance the competition. This is the purpose of the Just-In-Time Self Test.

Just-In-Time Action Plan—Master Planning

List items you discovered in this chapter that should be added to your JIT action plan:

1. _____
2. _____
3. _____
4. _____
5. _____
6. _____
7. _____
8. _____
9. _____
10. _____
11. _____
12. _____
13. _____
14. _____
15. _____

Chapter Ten

Purchasing

One of the most important groups that *must* support Just-In-Time flow is the purchasing department. Success for purchasing lies in the way it manages two things: organizing demand data for the suppliers and establishing solid relationships with the suppliers—its partners. Research shows substantial purchasing behavior differences exist between JIT and non-JIT companies. The following chart shows the results of an analysis of 18 companies, some JIT and some non-JIT firms, done by James Perry in 1988.

Element	Just-In-Time	Non-Just-In-Time
1. Vendor relationships	• Fewer suppliers • Longer contracts • Supplier partnerships	• Many suppliers • Short contracts • Adversarial relations
2. Supplier lead times	• Short delivery, 40 days • Shared data • Traffic planned	• Long delivery, 150 days • No shared data • Traffic unplanned
3. Purchasing methods	• Multiyear contracts • Short paper processes • Well organized	• Purchase price focus • Long paper processes • Inflexible methods
4. Inventories	• Reduced 10 to 50 percent • Improved turnover • Low finished goods	• Order point systems • Low turnover • Buffer stocks all over
5. Traffic management	• Freight cost trade-offs • Point of use delivery • Mixed small lot loads	• Excess inventory space • Costly receiving group • Large lot truck loads

These five elements are common to most JIT purchasing results. But they take time and effort. The questions in this chapter will help identify the issues to address to improve your Just-In-Time purchasing performance.

Chapter 10/Purchasing

Question 45. "Key volume suppliers are local."

Lead time from the supplier to the plant is a key issue. The closer the better. If daily deliveries are needed, the freight issues are minimized when suppliers are local. Milk runs and consolidated loads can be done in one day. But long distance travel is more costly and takes longer. This may mean creating a buffer stock to help offset the freight cost and act as a lead time buffer relative to how far away the supplier is. A better solution than setting buffer stocks is to use a mixed mode transportation process. Contract carriers can make milk runs in local or distant areas and bring the mixed loads to be consolidated into a trailer at the terminal. The multiproduct load trailer is then put on a "piggy-back" train and sent to your city where a truck brings it to your plant.

Question 46. "Single source suppliers make up greater than 50 percent of all suppliers."

A single source supplier is a supplier selected from a number of other suppliers. It is the one to whom a company expects to give all of its business for a part or set of parts. A sole source, on the other hand, is the only source available for a specific part or parts. Sole sources are often unavoidable, so they must be nurtured and carefully brought into a partnering relationship whenever possible. Single sources are easier to obtain because there is more than one to choose from and the probability of getting one that would perform well is much greater. Single source suppliers get all of the volume, thus generating better prices, more commitment, and greater loyalty. The quality guru, Dr. Deming, suggests it is almost impossible to successfully develop more than one high quality supplier.

Single source suppliers play a key power role for such companies as General Motors, Ford, Xerox, Motorola, and Toyota. They become partners in closely knit families and trade associations and are heavily dependent on their customers. Characteristics that exist in these partner relationships are:

- Most suppliers are geographically close.
- At some, 80 percent of all parts can be delivered within 30 minutes of need.

- There is enormous trust in the suppliers.
- Quality standards are not imposed by the customer but are set by the supplier.
- The suppliers have exhibited long-term quality performance.
- The suppliers are involved in the initial design stages of new products.

WHAT IS PARTNERING?

It is a permanent relationship that operates as if you and your supplier were in the same company. Partnering is a willingness to work together and to share savings and benefits. It is a realization that more is accomplished through harmony than through adversity. It is a mutual, ongoing relationship involving a commitment over an extended time and a sharing of information, risks, and rewards of the relationship. Often the focus is on the mutual elimination of process to process waste. Definitely a win-win situation.

WHY IS IT HAPPENING?

Partnering is occurring more today than ever before, because of the recognition that consumers expect greater quality and lower prices than ever before. Quality is desired more than price. As much as 50 percent of the quality failure of one company's output can be traced to supplier quality failure. It's also a buyer's market. So, the most effective way to eliminate quality failure throughout the supply chain is to form partnerships for mutual improvement.

Globalization of competition is also a driver behind the partnerships. Currently there is a worldwide surplus of capacity. This means that others have the ability to compete with us and that we can gain better market penetration with the use of offshore partners.

Proliferation and specialization of technological expertise drives some companies into partnerships wherein they must have certain technologies that are not able to be developed internally.

Question 47. "The number of active suppliers has been reduced substantially—by 50 percent or more."

This question acts in conjunction with question 46, but it is another measure of how well the process is working. As work is moving toward supplier certification, the number of active suppliers will reduce automatically. In question 19, some examples of reduction of supplier showed that as much as 90 percent of all products purchased were from only 10 percent of the suppliers.

There are some instances where suppliers are not willing to participate in partnerships or certification programs, so more than one may be required to maintain a reliable supply of parts with reasonable quality. But, the cost of receiving, inspection, accounts payable, purchasing, and stock keeping increases dramatically as the number of supplier are increased.

Question 48. "Buyers and suppliers are rated by supplier's quality, delivery, and ongoing improvement."

The role of the buyer has changed. In a Just-In-Time mode, buyers would no longer spend time on supplier selection; they would spend time helping the supplier to improve its processes. They would be coordinating changes of product design and schedules in cooperation with engineering and master scheduling personnel. They will be spending time to learn about the supplier's operations, quality systems, markets, and process changes. They will be helping the supplier to implement cost reduction projects from waste reduction plans.

Buyers should be rated on the same basic criteria that the suppliers are rated on. The buyer is the production manager for the supplier, just as the internal production manager is the manager of internal production. Thus, the measure for performance should be the same. The following are some measures to use to evaluate buyers:

1. Number of suppliers reduced.
2. Aggregate quality of supplier's receipts.
3. Aggregate number of late deliveries.

4. Number of suppliers certified and awarded.
5. Number of supplier surveys completed.
6. Percent of products versus number of suppliers.
7. Number of price reductions based on supplier improvement.
8. Net aggregate materials cost reduction.
9. Number of single source suppliers.
10. Number of supplier visits, face-to-face meetings.

All measures for purchasing should have a target and tolerance established as defined in Chapter Twelve.

Question 49. "Frequent multidepartment contacts are made between suppliers and your plant."

Traditional rules for buyer–supplier relations have generally prevented the supplier from having open contact with others in the buyer's company. Usually, the fear of others doing purchasing's job kept the contacts limited. If the supplier is to be a single trusted source acting as a potential partner, the relationship must be defined in a mutually beneficial and agreeable way. When the supplier needs data she may have free access with anyone in the buyer's company. However, when she needs authority to commit to a delivery, she must get that from purchasing. Guidelines must be established that treat the supplier as if her company were inside the buyer's company guided by internal policies and procedures. The following example of a full partnership relationship may help illustrate this point.

Hamilton-Hallmark, a large electronics distributor, has several customers that represent substantial portions of its business. It has been selected as the single source for all stockable electronic components that its customers use. One of the major customers, has entered into a simple partnership agreement that has the the following key elements:

1. The supplier will provide a full-time order administrator to be on site at the customer to manage orders between customer and supplier.
2. The customer will provide an office that can be secured.
3. The order administrator will have access to all necessary com-

pany records to facilitate the management of orders for the customer. He/she may attend planning meetings regarding requirements, inventory meetings regarding supply problems, and have access to the customer's computer systems in order to analyze inventory levels, shortages, and other purchasing issues.

4. The supplier will supply a computer terminal and modem connected directly to the suppliers data systems, whereby the order administrator may gather and report current status on parts and delivery to the customer.

The person assigned to one of their key customers was a bright energetic person with excellent analytical skills and good interpersonal skills. She fit in very well with the whole company. She went to daily purchasing or production meetings, reported on shortages, helped the purchasing department process its orders, investigated and quoted on parts for engineering's new products, and provided a personal link into Hamilton's management structure to lobby for special needs the customer requested. She acted more like an employee of the customer than an employee of Hamilton, although she was monitored very effectively by her regional management. She was highly autonomous and has been located at the customer's workplace for several years. She manages several thousand parts on a monthly basis. If she were to leave, it would probably require at least two buyers to fill the void with less ability to communicate effectively with Hamilton. This is truly a world class partnership.

>Question 50. "Most paperwork, material handling, tranportation, and quality waste has been eliminated between suppliers and plant."

Waste means anything that does not add real value to the product or service provided. One of the major waste creating processes in a company is the purchasing–receiving–accounts payable cycle. The real objective between the suppliers and your company should be a "last touch–next touch" concept for *waste free* delivery. That means the last person to touch the parts was the supplier's production person, and the next person to touch the parts will be the customer's production person. There is considerable waste or nonvalue-added activity that exists between the supplier and the customer's production lines. The following is a list of activities or paperwork that can be eliminated from the supplier networks as they become fully operative under Just-In-Time:

1. Loading and unloading.
2. Sorting parts
3. Moving or handling parts.
4. Receiving inspection.
5. Rework.
6. Expediting.
7. Returns to vendor.
8. Scrap.
9. Storage.
10. Repackaging.
11. Counting.
12. Purchase orders.
13. Acknowledgements.
14. Accounts payable posting.
15. Receiving reports.

If the parts are 100 percent quality, delivered on time, moved to the users by the driver, then items 1, 2, 3, 4, 5, 6, 7, 8, and 9 go away. If the parts are packaged in standard containers configured for exact quantities, then items 10 and 11 go away. If a signaling system like a kanban card, empty container or kanban square were used to define the next set of requirements, then items 12 and 13 go away. If the suppliers agreed to send only monthly statements for deliveries, then item 14 would go away. If a simple two-part slip were picked up by the driver for each container or kanban signal, then it could be used as a sales order for the supplier, a receiving slip for the customer (signed by a floor lead when the parts are brought in by the driver), and a data entry slip to update purchasing and accounts payable, thus eliminating item 15.

Much of the transmission of data and associated transactions can be eliminated by the use of Electronic Data Interchange (EDI). Many of the continuous and repetitive activities and documents like purchase orders, kanban cards, release signals to suppliers, and sales orders can be transmitted from your computer to your supplier's computer without any paperwork at all.

> Question 51. "Delivery lead time for most parts range from one day to one week."

This is a challenge. The shorter lead times means greater customer satisfaction. Often, customers have rapidly changing product life cycles or variable demands. Rapid change is necessary to keep them happy. Change is directly related to lead time. If a company can reduce its lead time internally to satisfy its customer demand, then the suppliers will have to be able to adjust accordingly. Thus, the focus for purchasing is to reduce the quoted delivery time from the supplier base.

Remember, the key to full Just-In-Time success is good supplier partnerships. Based on this partnering of suppliers where communication, coordination, and commitment are well developed, shorter lead times are attainable. There are several ways to do this with suppliers:

- Long-range schedules.

Longer-range production requirements provide the suppliers a clear vision of future demand. Whenever multiple products are supplied from the same family of parts, some flexibility could be requested for changes. However, the real issue is shorter lead times. Offer to help the supplier to identify constraints in the flow that could be removed. This creates shorter cycle times.

- Small lot production.

Large lot production takes longer to run and adds lead time. Smaller lots take less time to run and allow more different parts to be made. This means more setups in their processes.

- Setup reduction.

Setup reduction allows for faster runs and more different parts to be run. By analyzing setups they can reduce setup time by at least 50 percent without changing any tooling. They can do it by just cleaning up the setup procedures and by making sure that all setup activities that can be done while the machine is running are implemented.

- Contracting for capacity.

The greatest amount of time that a part takes to run in many suppliers' plants is in waiting to be worked on. Often the wait is 70 percent or more of the whole time a part is in the shop. To overcome this waste of time, buyers can negotiate the purchase of capacity, rather than a specific part. They would contract for let's say 50 hours of production time in the supplier's production line. The supplier would then release a "dummy order" representative of the typical run time of a part. Then when the buyer (or signal) releases the specific part to be made, the time is available and can be inserted in the line at the last moment. Obviously, the raw materials will have to be available. This process works best for product families where common raw materials exist. If the capacity is not used, the supplier bills the buyer for possibly 20 percent of the cost of the product they would have run, and then sells that capacity to another customer.

- Frequent deliveries.

Consistent, daily, or weekly deliveries of small lots can shorten lead times dramatically. If large lots are ordered monthly, the lead time is a month long. As long as the parts are repetitive and delivered daily or weekly, the only concern would be if the demand changes beyond the supplier's lead time to adjust to new higher volumes. Transportation system pricing is generally based on large volume or on full truck loads. Of course, larger shipments mean longer times between deliveries, tie up more space, cost more money, and require more people to maintain the materials. Smaller lots need less space, money, and people to manage them.

- Lead time waste reduction.

The elimination of lead time waste is easy to do. Flow chart the process from the last step in the supplier's production all the way to the first user of the parts in your company's production. Identify all steps that are not really necessary, or that can be systematically removed or reduced, and implement those reductions.

Just-In-Time Action Plan—Purchasing

List items you discovered in this chapter that should be added to your JIT action plan:

1. _____
2. _____
3. _____
4. _____
5. _____
6. _____
7. _____
8. _____
9. _____
10. _____
11. _____
12. _____
13. _____
14. _____
15. _____

Chapter Eleven

Data Integrity

Everyone in manufacturing needs to understand the impact of data integrity on the success of a Just-In-Time implementation. What is the accuracy of your current data? This chapter will show you how to create and report accurate and timely data. Examples and references will be provided for you to review. At the completion of this chapter, you will have reviewed or learned the following quality concepts and applications:

1. Inventory accuracy and transaction controls.
2. The importance of scrap and process errors.
3. How to set up and implement a inventory cycle counting program.
4. The impact of the bills of material on JIT.
5. How to structure bills of material for Just-In-Time systems.
6. How to define and use process routers for JIT flows and costs.
7. How to define work centers for JIT cost collection and standards.
8. How "process costing" works for JIT systems.
9. The management of process and product changes through engineering.

Again, be sure to keep a list of the detailed items in this chapter that you find to be missing in their own operations and add them to the JIT Action Plan. And check for other readings that will directly apply to your implementation.

Question 52. "Inventory record accuracy is 98-plus percent or better for both stockrooms and point of use storage."

The primary *technical* reason for Just-In-Time system failure is inaccurate records. This means that parts will not be in the right place at the right time if there are errors in the inventory records. All planning and execution of the planning systems and inventory backflush transactions are based on the correct quantities on hand. Errors, oversights, lost parts, and lots of other human errors occur that effect data integrity. Onhand inventory is the basis of planning requirements for supplier schedules. Though the JIT pull systems call for parts directly from suppliers, the suppliers must still have a long-term view of demand so they can plan their supplies and production scheduling.

Inventory accuracy monitoring is a daily requirement for JIT operations. At the end of each shift or day, a count should be made of all parts left in stock that were used that day. This may require stopping the lines a few minutes early each shift to allow for counting the inventory. This activity may be cycled to different groups or lines and spread out over a week's time. The objective is to catch errors in scrap reporting, shrinkage, and bill of material errors. This audit process is very important, because one part shortage caused by an error can shut down the line. Shortages become highly visible.

Visibility of part shortages are very high in JIT plants. An example of this phenomenon occurred at one the divisions of McDonnell Douglas Computer Systems Company. Many production personnel were asked what they thought was the accuracy of inventory. None of their estimates were higher than 92 percent accurate. After a thorough examination of the practices and policies for inventory control and accuracy, it was determined that the inventory accuracy was at least 99.6 percent accurate. But, because of the high visibility of shortages, the employees thought that low inventory accuracy was the problem. In truth, because inventory is so low, any error will cause a shortage even if it is a small error, as compared to the old ways where there was a lot of inventory on hand to cover these errors.

Question 53. "Bills of material accuracy 99-plus percent or better for costing needs and post deduct inventory."

A post deduct or backflushing function described in question 59 replaces all manual or frontend materials transactions from stock to work order. The elimination of these transactions uses the bills of materials to deduct the materials from stock after the work is completed. This eliminates the frontend issue transaction of inventory to jobs. If a product is completed, it has to have all the parts in it, so the bill of material, if accurate, can be used to deduct the parts automatically when completed.

> Question 54. "Bills of materials are flattened—structured with two or less levels in production."

Bills of materials (BOMs) are the core of the Just-In-Time data management. Almost all operations events are related to them. The bill of material is integral to all key functions of the organization.

Each function uses the BOM for a different purpose. Finance uses it for budgeting and pricing. Cost accounting uses it to collect manufacturing costs. Manufacturing uses it to track the production processes. Material planning uses it to plan materials to support production, and purchasing uses it to order the proper parts. Inventory control uses it to deduct and issue parts to production. Engineering uses it to define the parts, the processes, and the standards that are to be used to build the product.

It is obvious, then, as the core element of the JIT data system, the BOM must be accurate. There are several ways to maintain accuracy of the bills. One of the easiest is by checking the number of inventory errors that are found in the daily audits of inventory. Others are checking a completed product to verify that all the parts are there, compared to the bill of material.

One of the major changes from the former ways of production planning and control is that of using the bill of material for releasing jobs to the shop floor. Figure 11-1 shows the comparison between a multilevel BOM and a JIT BOM.

Each level in the bill of material, or single level bill, requires a separate work order to make it in manufacturing. Between each level, then, each completed work order is put back in stock to await the next higher level to be built.

In a JIT flattened BOM shown in Figure 11-1, all levels are

FIGURE 11-1
Bills of Material—Typical versus JIT

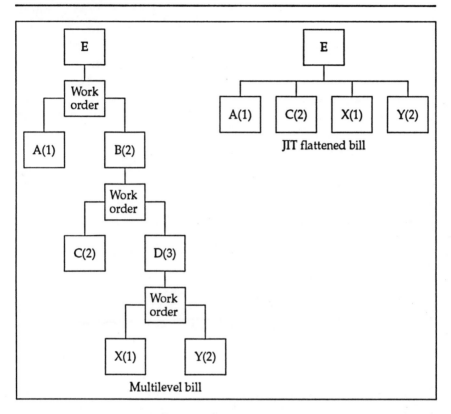

reduced to one or two. The routers are changed to include all the operations at one level in one process. Then only one "work order" would be required and *no* inventory between operations. As the pull system pulls the parts forward, *only* the necessary raw material parts are pulled from stock. All fabrication and assemblies are made in production only as needed, thus eliminating the need for stocking midlevel parts or assemblies.

The flattening process can occur in two ways. The first is to manually restructure the BOMs by taking out all the levels so the final parent has all of the components on one level. The other way is to use a "phantom" bill of material.

Phantom bills are simply done by a coding process in the computer that artificially removes the levels in the bill without having to do all the manual work. This process eliminates the requirement for a work order at each level identified as a phantom. A code is entered into the item or parts master file in the computer, typically a "p" or "x" or a number. This code tells the system to treat the assembly as a phantom. Phantoms are assemblies that are created *and* consumed inside the same work flow. Assemblies cannot be overbuilt and returned to the stockroom. Because they are still seen as components, they would have to be disassembled to put be put back in stock. If the assemblies are put in stock, there is no way the assembly can be pulled for a future use, because the part number won't show up as an assembly. Some systems have the ability to first determine if the phantom assembly shows an onhand balance, and, if it has, it issues the assembly and, if not, it issues the components.

With the use of phantoms or restructured bills, the Just-In-Time system will be faster, with less wasted paperwork, less confusion, less inventory, and increased throughput.

Question 55. "Routing detail, methods, or assembly instructions are accurately defined and maintained by timely process flow changes."

Routings describe the process steps used to produce a product. JIT continuous improvement changes the routings often. Flow changes common and dynamic and are often driven by self-directed work groups who are doing the work itself. So the methods updates will always tend to lag behind the process changes. This means establishing a highly responsive documentation process and CAD systems to update and print out new instructions, routers, and methods books. Obviously, this is a big task and additional staffing may be required. Unfortunately, the cost standards are affected every time changes occur to the flow.

A highly responsive engineering change order system is required to keep up with all the changes generated from the floor. Many companies have developed streamlined systems for managing timely engineering change requests and orders (ECOs).

ECOs are easy to implement in a JIT system. When it's time for the change, simply pull the kanban card or bin from the process and replace it with the new part; or, when a kanban card is sent to stores, purchasing, or supplier, change it then. Usually, there is little runout inventory to be concerned with, so changes can be done very quickly. If the kanban quantities are for only a few days or weeks of inventory, there is little obsoleted inventories generated, also. If there is no stockroom, the change is done.

When the supplier receives the next replenishment signal this can be flagged with the change and can be incorporated without the fear of generating out of revision stock because of late change notices and long lead times of large lots.

ECOs should be carefully coordinated with material handling, production, and any others who will be directing the flow of the changed materials. The materials could get misplaced because it will be a small lot. It could be missed by the production workers, or a part shortage may be recorded. Thus, it is imperative that the change be clearly communicated to production, material handling, and any stockroom or receiving personnel.

> Question 56. "Clearly defined engineering standards for costing."

Accurate engineering time or labor standards are required for the cost accounting process that identifies production costs by product or line. As noted in question 58, labor is "postdeducted" after the completion of each product by using the *labor standards* for each operation and comparing them to the total *actual labor charged* in all workstations used to make the products during the shift. The labor standards in the routings of all products must be as accurate as possible so analysis can be made to facilitate process improvement. Some companies have experienced as much as a 95 percent reduction in the number of shop floor transactions because the need to transact parts and labor movement are no longer necessary.

> Question 57. "Shipment forecast variation is ±10 percent or less by product family in the current time period."

One of the most common problems in American manufacturing

is poor frontend demand management, particularly shipment forecasting. In Chapter Nine we discussed the issues of master planning with a focus on accurate demand. The forecast of customer shipments is critical to effective Just-In-Time manufacturing. It is difficult to get an accurate forecast because the customer, market conditions, and quality issues are always there. But every effort should be used to get a forecast that will vary within a ±10 percent variance.

Customer shipment (not orders received) demand is hard to predict, but some techniques can be used to offset the forecast error. McDonnell Douglas Computer Systems Company had a problem with "jammed in" last minute orders from its customers. The company was always trying to rush an extra order through the Just-In-Time system it was implementing. It knew from experience that this was a historical trend and it wouldn't be able to change its sales people or the customers, and besides, the market was weak and it needed to produce everything that it could. In meetings among the manufacturing manager, master schedulers, materials management, and marketing, a decision was made to initiate a "P" order—a potential or projected order.

This "P" order would be scheduled at approximately the rate or times that the typical "jam-it-through" ("JIT") orders would have occurred. This enabled the company to plan the order as a regular product even without a sales order attached as a driver. Remember, in the pull system the sales order pulls the products from finished goods, which pulls from final assembly, and so on all the way upstream to the suppliers. So to plan the right balance of capacity and raw materials, an accurate volume of orders has to be known far enough in advance to be able to make parts and people available for the pull.

Check the reference for focus forecasting in the Bibliography.

> Question 58. "Accounting systems and controls have been redesigned to work in a JIT environment."

Successful Just-In-Time manufacturing requires continuous improvement. Continuous improvement causes changes to traditional procedures. Cost accounting systems are very traditional. They don't change easily. Cost accounting is the function of the overall accounting system that collects costs from operations (essentially labor, material, and overhead) and totals them by product. These totals are then combined to provide the overall cost of making all products or *cost of goods sold*. This information is then used as the base for pricing the products for the marketplace.

Traditional cost accounting systems focus on *control*. That is, the data is collected after the fact and is used to redirect the processes to bring costs under control to keep the customer's price down and sales up. Thus, by collecting detailed data on each operation, process, product, or component, the analyst should be able to pinpoint which particular location in the manufacturing processes has created excess cost. Once the excess cost is identified, the theory is that people can act to control this point to bring the cost back under control.

The deficiency of this approach is that the control of cost by element can seldom be timely and that cost variances continue to change as processes change. In a continuous improvement environment, changes are the norm. So cost accounting data will usually be different every time collected. Consequently, there is little control value to the data. Also, the masses of detailed data become very difficult to analyze, and they are usually made available after month-end close, which is too late to find and fix any cost variances. The solution is to redirect and redefine the purpose of cost collection. It must go deeper, looking at the real causes of variation in the processes and then deciding what cost data should be used.

The new criteria for cost accounting systems are flexibility and quality. Just-In-Time cost accounting systems will focus on significant events, rather than pure cost transactions. Data are collected on *cost drivers*. As processes become more flexible and as quality improves through improvements to the process, the system must reflect the relevant cost changes that actually impact the overall process costs.

Cost collection is done by operation or process, rather than by product—that is, there are *no work orders*. Instead, cost is collected from each operation that *drives* or contributes an added value to the product. The following is a comparison of the differences between job (work order) cost and process cost.

Job costing uses cost data determined before production and compared to actual costs to determine variances.

Process costing uses data collected by time periods. Costs are averaged across all units during a period.

Job costing uses work orders to collect actual cost data.

Process costing uses a work center, or cell, to collect actual cost data.

Job costing collects data during production.

Process costing collects actuals after production by "post deduct" or "backflush."

Essentially, process costs for a part remain the same unless something changes, like scrap and rework. These are the only manual or exception transactions actually needed. All other data would reasonably be the same. As automation increases, less and less labor will be required to produce the product, so labor costs will become a part of the overhead or burden of the workstations and no longer be tracked separately as a cost driver. It will join the other cost drivers, like power, engineering support, material handling, quality auditing, and other "indirect" drivers directly relevant to the workstation or processing center.

The backflush or postdeduct function replaces two major sets of transactions—materials transactions from stock to work order and labor data to work orders. The elimination of these two typically manual sets of transactions has resulted in greater than 75 percent reduction in the number of transactions. Backflushing means using the bills of materials to deduct the materials from stock after the work is completed. As referenced in question 56, this eliminates the frontend issue transaction of inventory to jobs. If a product is completed it has to have all the materials in it, so the bill of material, if accurate, can be used to deduct the parts automatically when completed. This then leads to the elimination of a separate stockroom transaction, because inventory will not need to be separated for control purposes but be brought directly to the production floor in small quantities. Parts are used on an as-needed basis, thus only those used for production will be deducted.

A new category of inventory has been created in many companies called "raw in process." This inventory can be created by two mechanisms: either through a location to location transfer from raw material to raw in process, or directly from receiving (purchase order) to raw in process. Visualize this picture: pick up one of your production lines and place it in the main aisle of the warehouse. Have the workers take whatever they need directly from the racks, have the materials located on the racks and to make it easy to reach. At the end of the shift, deduct only what was used based on the bill of material of what was produced. Remember to record all scrap carefully and deduct it at the end of the shift, also. This is how the raw in process concept works. Other than scrap, there are no manual transactions—thus eliminating hundreds or maybe thousands of transactions daily.

Labor tracking is also simplified in a similar manner. Instead of having the employees post their labor hours to each operation of each job worked on each day, they will only clock into each workstation where they work each day. This eliminates almost all of the daily transactions done by hand. A recent report from an electronic assembly house showed it eliminated 28,000 transactions per month by going to the JIT process. Visualize this: make a list of all employees working on a particular assembly line. Give it to the line supervisor and have her record the start and stop time of each person working that shift as well those moved to and from other lines. Turn in the list each day and total all the labor on the sheet. Post and spread the totals to all products worked on that line that day. Compare these actuals to the totals generated by the labor standards in the routings of the products produced that day. Any variances will be in aggregate, but they will tell you if the overall process was productive *that day*. Automated computer systems are available to do this work.

Just-In-Time process costing focuses on the overall optimization, rather than suboptimization of individual processes that are very difficult to fully integrate into a relevant picture of production costs.

Just-In-Time Action Plan—Data Integrity

List items you discovered in this chapter that should be added to your JIT action plan:

1. _____
2. _____
3. _____
4. _____
5. _____
6. _____
7. _____
8. _____
9. _____
10. _____
11. _____
12. _____
13. _____
14. _____
15. _____

Chapter Twelve

Results

The last and maybe most important area of concern in your Just-In-Time journey is that of measurement and reporting in a JIT environment. Compare what you report now to the list of typical JIT reports in this chapter. You will be shown the proper measurement tools to use and when to use them. Examples and references will be provided for you to review. At the completion of this chapter, you will have reviewed or learned the following quality concepts and applications:

1. The importance of performance visibility.
2. How to track daily, weekly, and monthly with key indicators.
3. Promoting results to the organization.
4. Typical results from accomplished Just-In-Time manufacturers.

Be sure to keep a list of the detailed items in this chapter that you find to be missing in your own operations and add them to the JIT Action Plan at the end of the chapter.

> Question 59. "Operational measurements with both targets and tolerances are in place and reviewed daily by management and employees."

If the instrument panel in the dash board of your car went blank, you would probably still be able to drive for a while. But you wouldn't be able to tell how fast you are going, how far you've gone, or how much gas you have left. Imagine an airline pilot without an instrument panel. Then think about the many complexities of a production operation. If measures were implemented, they would look more like an airplane's panel than a car's dash

FIGURE 12–1
JIT Measure Board

Master Schedule Performance	Units per Production Hour	Units per Payroll Dollar	Cost of Quality	TQM Team Performance	Absenteeism
Schedule Performance TRADITIONAL	Overall Cycle Time	Inventory Levels	Supplier Quality	Quality Performance Subassembly	Accidents per No. of Employees
Schedule Performance FAB	User on Time Delivery	Inventory Turns	Shortages	Quality Performance TRADITIONAL	Quality Performance FAB
Schedule Performance ASSEMBLY	Schedule Performance MAG	Inventory Accuracy	Field Returns and Repair	Quality Performance MAG	Quality Performance Assembly
Schedule Performance 5000M-2	Supplier Delivery	Cycle Time Assembly	Floor Space Reduction	Quality Performance 5000M-2	Quality Performance FINAL TEST

board. In a business, there are many drivers besides the president. The president needs to know the speed and direction in order to take action. The objective of measurements is to show how the vessel is performing against a plan. Everyone must be able to see the plan as well as be guided by it. That plan might be called a Just-In-Time "measure board." The following sample shown in Figure 12–1 is an example of several JIT measure boards installed in JIT companies.

Note the several categories on the JIT measure board: schedule performance, quality performance, cycle time reduction, supplier performance, delivery performance, team performance, and accuracies. Each of these measures is selected to show the overall quality and delivery parameters of the organization. But others can be added that show other desirable parameters, such as absenteeism, accidents, productivity in volume per employee, and qualitative performance measures.

Each measure must have a target line and a tolerance line. Target lines are established by analyzing each of the key components of the measure. For example, cycle time could be measured in many ways—from the very first action to pull the part in a stockroom, or from the first action on the shop floor; to the time it hits shipping or the time it goes on a truck to the customer. One company chose to use the time the inventory was first pulled to the time it went on the truck.

Once the definition of the measure is set, a target can be determined. Targets should be driven by company policy. For example, if the cycle time was 20 days to produce a product, and the company decides that customer delivery must be at 5 days to meet or beat the competition, then the target line on the chart would begin at 20 days and, over a period of months, as decided from a thorough analysis of issues and continuous improvement activities, the line is drawn to the 5-day point on the chart in the appropriate month to meet that goal. This same logic should be used to establish most of the measures.

Some measures are relatively easy to establish, like performance to schedule, on-time delivery to customer, and so on. But one should steer clear of efficiency measures that relate to any one group or department. These tend to suboptimize performance and focus efforts at localized activities. All measures should be tied to overall throughput for them to be relevant. For instance, if one department is measured on the number of parts per employee, but the output only builds up inventory in the next operation or in stock, it has only created waste. This "push" type measure does nothing for the overall customer delivery—it just builds inventory that costs in most companies around 2 percent per month just to have it. A "pull" system will only call for those parts that are needed, so the efficiency that overproduces inventory is less relevant than overall throughput to shipping.

The second key part of each measure is its tolerance. Tolerance is generally based on those things that cannot be controlled by those performing the work. For example, performance to schedule could drop in any one period because of a supplier quality failure, or when a hurricane kept the parts from arriving. Tolerances might be fairly loose at the beginning of the measurement and tighten over time. For instance, the tolerance for on-time supplier delivery at the beginning might be as wide as 10 days late; but, as inventory is reduced, suppliers become certified, smaller lots are instituted, and supplier quality gets better, the delivery tolerance should be narrowed to 1 or 2 days or less.

Another purpose of the tolerance is to identify, like an SPC control chart or run diagram, when trends are getting out of tolerance or limit. As processes are brought under control, the reading should always be within tolerance. But, where SPC control charts are used in operations, they are tracking continuous and repetitive performance hopefully without much change. If the process were constantly changing, the SPCs would have to be changing also. So run diagrams are more appropriate for the JIT measure board. A run diagram only defines an upper and lower *specification or tolerance* limit, rather than *process defined upper and lower control* limits. Some examples of these type of measures are shown in the following questions.

Question 60. "Overall manufacturing process cycle time and throughput lead time are reduced."

An example was given in question 59, which defined how a cycle time measure could be set. A typical mechanical method for tracking and reporting the cycle time would be to attach a dated tag or sticker on the very first part pulled or used to make the product. Then the tag or sticker is removed when it enters shipping. It would be logged and charted at the end of each day for each product shipped. All products would then be totaled by week and posted to the JIT measure board.

The results of continuous improvement activities focused on quality improvements will be seen in less rework, less scrap, less inventory, less waste, and less paperwork. Cycle time and lead time are directly affected by these results. An example of a JIT measure board chart is shown in Figure 12–2.

Question 61. "Production floor space has been substantially reduced."

The single best indicator of shorter flows, less inventory, and more integrated Just-In-Time processes is the freeing of floor space in production. Many companies put in recreation activities in these free areas. Of course, these areas can be used for expansion into new product lines. For example at—a plant in California was being moved to a new facility to expand for growth. The produc-

FIGURE 12-2
Supplier On-time Delivery

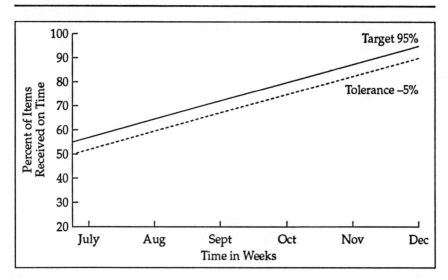

tion group thought it would get more space to spread out its operations to have "more room." The general manager, however saw this as the antithesis of good JIT flow and set the space requirements in the new facility at less than the space production currently occupied in the old building—reserving the extra space for expansion into new products in the future. At first, the supervisors were resistant, but, after the new floor layouts were completed, redefined, and flows improved again, they were easily able to do it. The exercise made them think carefully about their current inefficient use of space. Floor space reduction is measured as the amount of floor space freed, because the more space freed, the more production space has to shrink.

Question 62. "Work-in-process inventories are continuously reducing."

Work-in-process inventories are a function of at least two things—work orders that exist, and the lack of a tight pull system in the shop. In the pre-JIT environment, work orders are the only way to track products through the shop. They act as a bucket traveling through production collecting parts and labor. At monthend when the books are closed, whatever shows up as inventory in all of the "buckets" is the amount of work-in-process inventory. The JIT system will not be using work orders in this manner. There won't be any work orders—thus, no inventory. Of course, the inventory hasn't all gone away, but the method of tracking it has. As indicated in question 58, work-in-process inventory becomes "raw-in-process." That means the inventory never moves out of stock. Second, the inventory will only be used as needed on a pull basis, so no excess should be there.

Excesses will still exist; but, as lot sizes reduce, cycle times become more balanced, and quality improves, raw in process also will begin to reduce. However, at the beginning of Just-In-Time manufacturing, where product lines are converted one by one, raw in process will be increasing while raw material will be decreasing. Decreases in raw material stock will occur as more and more materials are brought directly from suppliers to raw in process—eliminating double handling and double transactions. Decreases also will happen because of smaller lot sizes from suppliers.

Thus, a measure of the reduction of work-in-process inventory should include raw-in-process inventory as well. Over time, this measure may become irrelevant as a meaningful measure as inventories approach a minimum.

Question 63. "Substantial increases in productivity or shipments per employee."

This is only one of several measures of the results of Just-In-Time systems. Sometimes this number is called "throughput." Throughput means that the products have moved directly to the customer, not to finished goods stock. So a productivity measure of unit volumes of parts shipped per employee means that the overall company has performed for the customer. Some companies like to measure productivity by parts produced to finished goods. This will distort the measure, because parts may be getting produced that are not shipping and the excess inventory becomes waste. It means that the employees produced something that wasn't needed.

One key indicator for financial purposes may be expressed in dollars shipped per employee, or as dollars shipped per labor hour. This is indicative of productivity.

This measure becomes an excellent sales and marketing aid. It shows the customers that continuous improvement efforts are working and that lead time will be much more predictable and shorter. For subsidiaries of larger corporations it is an excellent bottom line performance indicator.

Question 64. "Substantially reduced operating expense."

Operating expense is essentially tied to the amount of activity that goes on inside a company. As noted in question 58, one company eliminated 28,000 work order transactions per month just by eliminating the detailed shop floor transactions it had been doing for years. Several people were freed to fill other jobs, and less people were hired to support their growth.

Less inventory requires less people, storage areas, material handling equipment, and power. These are all elements of carrying costs of inventory. The typical carrying cost of inventory in the United States tends to be about 20 to 30 percent per year. The major components of carrying cost are:

- Obsolescence.
- Warehousing.
- Record keeping.
- Material handling.
- Interest or capital costs.
- Insurance.
- Taxes.
- Shrinkage.
- Scrap.

Reduction of inventory will increase the direct contribution to the bottom line through the reduction of operating expense, as shown in the example on page 142.

You'll notice an inventory reduction from $9 million to $6 million between the current plan and the proposed plan. Using the formula on the proposed plan, it shows that 18 million in cost of goods sold divided by 6 million is 3. That is, the inventory would be bought and sold about three times during the year versus only two times in the current plan. To determine the savings, take the carrying cost of 30 percent of the $3 million inventory reduced. It will equal $900,000 in savings from carrying cost reduction.

FIGURE 12-3
Cost Impact of Inventory Reduction

Formula: $\frac{\text{Cost of goods sold}}{\text{Average inventory}}$ = Inventory turns	Current Plan	Proposed Plan
Cost of goods sold	18M	18M
Average inventory	9M	6M
Inventory turns	2.0	3.0
Inventory reduction	0	3.0M
Inventory carrying cost	30%	30%
Savings from carrying cost	0	$900,000

Caution must be used here, however. The reduction in carrying cost means that the elements of carrying cost mentioned earlier must be reduced or removed from the company expenses. People must be removed and warehouse space subleased or used for expansion for new products or processes. Otherwise, the benefits from savings in carrying cost are a myth.

Other elements of operating expense will be reduced, as JIT is implemented. Quality failure—doing anything poorly throughout the company—can be as high as 30 percent of a company's operating expense. Consider all the effort expended to inspect, track, handle, rework, move, store, meet on, report on, and manage faulty processes or products throughout the company.

Question 65. "Overall cost of quality is continually reducing."

This is related to question 64, in that all of the cost of quality that can be recorded is tracked. Tracking the cost of quality can be done by segmenting the costs into at least the four categories—prevention, appraisal, and internal and external costs. From this, a strategy can be developed. A reasonable strategy would be to reduce and eliminate the bottom two categories, internal and external failure costs by moving the costs, up to the prevention category.

The measurement data can be collected from departments and posted to specific account numbers for tracking, budgeting, and analysis purposes. Often these costs are collected but never reported in this manner. They are usually reported as operating expense. The strategy for improving quality must be based on an approach like this. If there is a declining monthly target set by the company management starting at $250, 000, and a tolerance of +$20,000, then continuous improvement activity can be measured. The chart should be posted monthly.

Question 66. "Stockroom inventories are continuously reducing."

Stockroom inventories are usually reduced through smaller lot sizes and because they are being transferred to a "raw-in-process"(RIP) type category, which is a location. As defined in question 58, RIP inventory is directly available for production and is decremented or reduced by the backflush transaction at the end of the shift or day. This is a simple chart to be developed. It is usually incorporated in an overall inventory chart showing each inventory classification, such as the example in Figure 12–4. Notice that the trends of inventory are continuously down. That's the plan!

Question 67. "Substantially increased inventory turns."

Inventory turns is a typical measure that has been used for years. There are several variations of it. The example here will be based on the simple formula: cost of goods divided by inventory. The cost of goods is the total of materials, labor, and overhead and/or burden costs. Turns represent the number of times the inventory cost in dollars has been replaced or turned over in the company in one year. So four turns means that inventory was bought and sold four times annually, or about once every three months. Profit or "mark up" is added on top of the cost of goods to set the retail price for the goods to the customers. Consider this, if inventory turned one time per year, that would mean there was

FIGURE 12-4
Inventory Levels

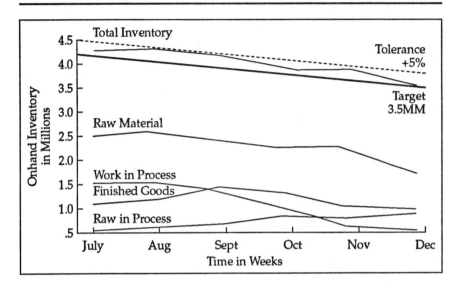

one profit amount earned from the sale of the goods. If, however, four turns occurred, this generally means four amounts of profit were taken. So the more inventory turns through the company, the more profit "hits" will occur.

In the example in Figure 12-3, question 64, a company proposes to reduce inventory by $3 million. The number of turns based on the formula goes up from two turns to three turns per year. The bottom line savings that could occur if the carrying cost of inventory was 30 percent is $900,000. There are a few caveats about this savings that must be highlighted.

Carrying cost is composed of several items associated with the inventory, such as warehousing, material handling, scrap, shrinkage, obsolescence, deterioration, taxes, insurance, and finally, the largest number—interest expense. Interest expense, or the cost of capital, is the cost of having to pay interest to a bank for money borrowed that wouldn't have been borrowed if the money wasn't tied up in inventory. Inventory reduction frees cash for other investments. Thus, inventory turns becomes a reasonable indicator of overall continuous improvement generated through all the actions taken to improve throughput, reduce operating expense and, reduce inventory.

> Question 68. "On-time customer delivery is continuously increasing for both line fill and order fill rate."

On-time customer delivery is probably the most meaningful measure any company can use. It is the final proof of performance. It is shocking that so few companies in America measure their customer delivery percentage. There are several ways to measure on-time customer delivery. One way is by order fill rate. Simply divide the number of orders filled by the total orders to be shipped per day, week, or month and multiply by 100 to get a percentage. It would be best to do it by week to show the weekly trends that occur during the month. This approach would show up month-end "crunch" problems, wherein a major segment of the monthly requirements are shipped in the last few days of the month to meet a monthly financial target, rather than a customer satisfaction target.

Line fill rate is a percentage derived from the number of line items filled per order per period for companies that ship multiple items on single orders.

Sometimes a filter or window is used for these measures by adding a variability adjustment. For example, if the measure was based on the number of orders shipped versus the number scheduled per day, a filter is added that modifies the "day" to, say, two days early to one day late. So any order that ships between two days early and one day late is considered as shipped on time. As you can see, this would become a "fudge factor" that would make the delivery percentage look better than it really is. However, there are times when unavoidable delays could occur with carrier schedules, customer changes, or weather that might prevent orders from shipping that were beyond the control of the company. Don't use this tool to cheat. Use it as a startup tool and tighten it as deliveries get better. Be sure to post the window definition on the chart so people will know it is being used.

All measures of customer delivery must be based on Just-In-

Time philosophy. That is, the company is supposed to deliver product to meet customer demand. If it is produced to put into stock instead of ship, then the system is not as sensitive to customer need. For example, if a "buffer" inventory was established for finished goods that is only intended to accommodate excessive customer demand in a make to stock company, that buffer needs to be as small as possible to eliminate excess inventory, and to exercise the Just-In-Time system's ability to respond quickly to customer needs. Caution must be used in setting these buffer stocks. They will surely become a crutch. As the lead time gets shorter and shorter for internal production, the finished goods buffer should be less and less.

Just-In-Time Action Plan—Results

List items you discovered in this chapter that should be added to your JIT action plan:

1. _____
2. _____
3. _____
4. _____
5. _____
6. _____
7. _____
8. _____
9. _____
10. _____
11. _____
12. _____
13. _____
14. _____
15. _____

IMPLEMENTATION

Chapter Thirteen

Building an Implementation Plan

The purpose of this chapter is to provide a technique and format to quickly and simply build a Just-In-Time implementation plan. You've been asked to record any items in each or the previous chapters in the Just-In-Time Action Plan forms. Now go back and collect the data from the previous eight chapters (Chapters Five through Twelve) and enter them into the outline form called the JIT Project Plan at the end of this chapter. This chapter will outline the rest of the elements that might be needed to set up a good JIT project plan.

START UP TEAM STRATEGY

There a many ways to organize a JIT project, and much has been published about how to do it. This section will describe one way to start.

1. Just-in-Time Leadership

As usual, top management has the responsibility for getting things going. If the managers are prepared to implement Just-In-Time and truly believe in it, of course it will make the job easier. But, most of the time this isn't true. Often they need to be convinced by subordinates that Just-In-Time is the right strategy. The best argument is in dollars. The chart in Figure 13–1 shows a model of bottom line impact of JIT implementations.

Figure 13–1 shows how two company model strategies might compare. The first strategy was to do as most companies do,

FIGURE 13-1
Bottom Line Impact—Just-In-Time

	Percent	Current year	10% Sales Increase	20% Materials Cost Reduction
Sales	100	6M	6.6M	6.0M
Materials	40	2.4M	2.64M	1.92M (20% Reduction)
Labor	30	1.8M	1.98M	1.80M
G&A	20	1.2M	1.32M	1.20M
Total Cost of Sales	90	5.4M	5.94M	4.92M
Gross Profit	10	0.6M	0.66M	1.08M
Added Profit			0.06M	0.42M
Percent Improvement in Profit: 80%			10%	18%

increase sales by 10 percent. In the current column, sales were at $6 million and, under the 10 percent sales increase column sales are posted at $6.6 million. Notice that materials, labor, and the associated general and administrative (G&A) expense (which in this example includes overhead) all increase by 10 percent. The added profit is $0.06 million or 10 percent.

In the second strategy, the company chooses to reduce the materials expense through a Just-In-Time implementation. Most of the reduction to materials expense comes from quality improvement, carrying cost reductions (normally seen in overhead), and purchase cost benefits derived from long-term contracts with suppliers. Notice then, that a 20 percent reduction in materials expense would contribute $0.42 million added profit, or an 80 percent improvement in profit (from 10 to 18). Understand, this is only a model. If an internal analysis were done based on your own profit and loss and balance sheets, looking at all the benefits derived from Just-In-Time the numbers would probably be startling. There are many examples throughout the JIT Self Test that would help the development of such a model.

A motivated top management team is a blessing. However, there will always be some resistance. Many of the "inhibitors" (revealed from case studies), which have prevented more rapid implementation of Just-In-Time manufacturing, are the following:

 a. Poor management support. Management support is generally seen in the actions of a steering committee. The following list shows the type of support needed from top management through the actions of the steering committee:

- Steering committee is headed by top management.
- Resolves policy issues as they arise from the project teams.
- The steering committee defines clear-cut objectives.
- Provides resources to support changes to process/people.
- Sets or approves priorities.
- Provides visible rewards and credit to teams.

 b. Rework as a "solution to failure." Old habits are hard to break. It is easier to rework a failure than to stop the line and leave everyone sitting or waiting or moving to other areas. Most workers are conditioned to keep producing regardless of someone else's problem.

 c. Reverting back to old planning habits. The most common and visible old planning habit is releasing or supporting larger batches in production. Somehow the mind seems to think that a larger lot is more "efficient," and it becomes too easy to run a larger lot than to constantly do setups and changeovers—especially when the worker knows that the same part will be running again in only a very short time. People don't like to change setups; it's harder until it becomes a habit.

 d. Quality assurance resistance to change. The first big change is the elimination of inspection from the processes. The next one is the elimination of receiving inspection. The fear of job loss arises and causes increased resistance to quality at the source. Their retraining entails learning how to facilitate others' performance, train others in quality techniques, spend more time on analysis and collection of process data, or work on supplier certification programs. These jobs were probably not in their original job descriptions.

e. Resistance to process flow changes. Considerable effort is spent by floor supervisors, lead persons, manufacturing engineers, and others to get a line "just right." Then it gets changed again and again and again. The resistance is there because people want things to be predictable. Constant change is very uncomfortable psychologically. The human mind always seeks the status quo. Often the resistance comes from the fact that the origination of the changes didn't include them.

f. Low support group involvement. Support groups like purchasing, engineering, sales, and accounting are not measured by the flow of material through the process, even though they have considerable impact on it. What happens to the flow of incoming materials if purchase orders don't get placed, or accounts payable doesn't pay on time, or engineering doesn't release a producible part, or cost accounting pushes for larger batches to reduce the apparent cost per piece. If these groups are not intimately involved with the process changes that invariably lead to policy changes, they will not be in support of the changes.

g. Limited education and training. One of the oldest rules in the industry is "Success is directly proportional to the amount of education and training applied." This rule applies to Just-In-Time more than most other projects, because JIT is more fraught with considerable changes to process *and* procedure throughout the organization. Limiting the amount and type of education and training is like limiting your success.

h. Uncontrollable economic/marketplace conditions. This circumstance is the easiest to justify in terms of resistance. But, it is well known that companies closer to the survival level are usually the most successful at Just-In-Time implementations, simply because they have to do something, and they have nothing to lose and little to protect. Job fear goes away because it is redirected to company survival instead of Just-In-Time changes. One company in a survival mode gave the employees a pep talk about the economic conditions as related to the Just-In-Time project. The employees were told about the JIT plan and two choices were given them: (1) they could refuse to support the Just-In-Time plan and lose their job anyway or (2) they could fully support the Just-In-Time plan and gain invaluable personal and professional growth. If the company succeeded they would all win. But, if they lost their job, they would be greatly valued by other companies and not have to fear unemployment.

2. Just-In-Time Team Formation

Teams are the way major projects get done today. Seldom does one person have the knowledge, insight, or volume of time necessary to accomplish complex tasks in the modern manufacturing world. Self-directed work teams are the answer. They are not easy to form and even harder to keep motivated. An organization for the team was presented in Chapter Five, question 8.

Project organization is crucial to successful implementation. The members of the project team are from each area of the company. *Note:* in question 8 they are called "sponsors." A sponsor is one who acts in support of the person doing the work. Each sponsor has a responsibility to act as a liaison between the groups that he or she sponsors and the JIT project team. He or she acts as a guide and auditor for them.

There are at least two other defined roles in each group, the group liaison and the group recorder. The purpose of each role is to spread as much of the responsibility as possible to the members of the group and to rotate them occasionally. When members have some small amount of responsibility, they will be more involved and supportive of the group purpose.

3. Mission Statement

The next task is to define a mission statement for each of the work groups. Mission statements for the teams should define the following:

- What the problem or area is.
- What the boundaries are.
- What is to be done.
- What the expectations or results will be.

Teams can get lost or go off track very easily. Some of the reasons why this happens is tied to the mission statement. Some of the problems with mission statements are:

- The statement is too broad.
- The project is beyond the group's capabilities.
- There is no mission statement.
- The nature and importance of the project is unclear or unknown.

A poor example of a team mission statement would be "Increase inventory accuracy" or "Clean up the shop." A good example would be "Analyze all inventory transactions and identify the most common errors and resolve them at the cause." Or, "Survey the shop floor and identify all materials, tools, and equipment not necessary. Remove or properly locate all of these elements."

4. Education and Training

If serious commitment to Just-In-Time exists, then a training group or department must be defined. There must be trainers selected, both internal and external. Train-the-trainer sessions should be given to develop teaching skills, and then a program of the basics should be started.

Again, a mission statement should be defined for the training group, such as "To enhance personal growth, professional capabilities, and technical competence of employees in concert with the overall goals of the company." The training curricula should be broken into at least two groups, *support group training* for accounting, marketing, engineering, purchasing, and so on, and *production group training* for materials management, inventory control, production operations, quality assurance, maintenance, and the like.

Education and training are a lead time problem. It is very difficult to get most of the company's employees through comprehensive training in a short time. Short focused sessions (one or two hours) balanced against production and quality needs should be scheduled. A set of basic courses to begin with could be:

- Just-In-Time theory.
- Total quality management.
- Problem solving.
- Facilitator training.
- Customer/supplier relationships.
- Just-In-Time working model.
- TQM tools and techniques.
- Statistical process control.
- Self-directed work groups.

So, if you only taught two hours per session for the above nine subjects, to an average group of 20 people per session, and for a population of 200 employees, the number of classes would be 90. The schedule, which should be designed not to pull more than 10 percent of the people from production daily, would be at least three months, if run every day, and all people were to attend. In truth, the schedule would take more like five to six months and is only for the basic subjects, consisting of two hours for each of the nine subjects. Each employee would get 18 hours of basic training.

Along with the overall education and training needed prior to the Just-In-Time implementation, JIT education can be used to overcome the "stagnation" that often occurs during an implementation. Conduct an evaluation of areas or disciplines that appear to be slipping. Solicit worker opinion regarding where and when to do training, and then design and conduct the training right on the shop floor, taking working examples right from the surroundings. Invite support groups and members from all areas, mix them evenly in each group session of about 20 to 25 people. Keep the attitudes positive, forgiving, and constructive. Prior to the first session, post a list of required attendees and ask if others would like to attend.

There are many others ways of obtaining knowledge in support of JIT manufacturing:

a. Reading current books on JIT. Establish a company library.
b. Visit plants in other companies implementing Just-In-Time.
c. Learn by doing the work in new ways.
d. Share experiences with others in group sessions
e. Join or attend professional societies like the American Production and Inventory Control Society or the American Society for Quality Control. These societies have dinner meetings monthly, local workshops, regional and national conferences.

5. Pilot Selection and Preparation

The steering committee selects a pilot line or area to begin with. Don't attempt to do it all at once companywide. Just do a pilot first. The pilot should have the following characteristics:

a. It must have a high probability of success.
b. It must have good skilled workers.
c. The workers are trained in JIT.
d. It can be started quickly.

After selecting the pilot, develop an implementation plan, which should include:

- Continuous quality improvement.
- Workplace cleanup and organization.
- Reducing setups.
- Reducing lot sizes.
- Reducing lead times.
- Process or flow layout changes.
- Convert from work order to pull system backflush.
- Focus on supplier quality certification.
- Deliver parts direct from supplier to pilot line.

6. Define the Just-In-Time Model

All of the knowledge gained from the pilot should have been recorded and prepared through training and documentation to the others in the plant who will be selected to bring up the next Just-In-Time sections. A model of each of the major techniques used and their impact should be defined for others to follow. Certainly, a list of those issues that were major problems must be prepared and presented along with the benefits achieved from the implementation of the pilot.

7. Keys to Success in Team Meeting Management

The failure of any meeting is the responsibility of the meeting leader. The following is a list of the elements of meetings that should be well managed:

   a. Distribute the agenda before the meeting.
   b. Start and end on time.
   c. No interruptions or side conversations.
   d. Use consensus decision making or two-thirds majority.
   e. Everyone is responsible for staying on track.
   f. Limit discussion to specific time period.
   g. Listen!!
   h. Allow flexibility in the agenda.
   i. Assign responsibility and due date for each action.
   j. Summarize issues and decisions at end.
   k. Assign responsibility and due date for each action.
   l. Keep and publish minutes. Rotate responsibility.

There is much more available than can be covered here on meeting management. Meetings are a major communication requirement. They are where assessment of progress and task assignment occur. All members in attendance are plugged into the communication network and must always be there. There must always be meeting minutes and meeting agendas.

Meeting agendas are required for every meeting. They provide a guideline, a time plan, and subject index and a distribution list. Attendees should be tracked and attendance reviewed by top management occasionally. A "decisions/policies made" section becomes a good reference for looking up previous decisions. This will be used a great deal, because people have short memories, and a lot of activity tends to cloud memory. Notice that actions must be assigned and dates set so all will know the tasks and dates of critical elements to be done.

8. Obstacles to Implementation

There will be many obstacles to the implementation process. Some of the following items have been noted by implementors and experts.

   a. Having adequate time to do the tasks along with other tasks. This is one of the most common issues. Because Just-In-Time is an ongoing improvement process, this issue will always exist. Time management becomes an important tool.

   b. Trying to do too much. The Just-In-Time process is an ongoing set of events; it doesn't have to be rushed. However, people are always excited to obtain the benefits in short term. They may not want to wait, so pressure is put on the team.

   c. Too narrow or too wide a scope. As addressed in the issues about mission statements in this chapter, pick a pilot or task that gives desired benefits without taking too much time or more resources than are available.

   d. Poor measurements. Many companies are notorious for not measuring themselves. For the JIT system to be visible, it must be properly measured and visibly reported, as noted in Chapter Twelve.

   e. New management roles. The greatest issue facing management during the conversion to JIT manufacturing, besides change itself, is the managers' resistance to give up control and to force decision making downward to the worker and then provide support for what gets decided. The second major problem is converting the manager from an autocrat to a facilitator.

   f. Unsustained progress. It is a common occurrence that progress comes in spurts. It is difficult to maintain *continuous* improvement. A review process by management will overcome this problem.

   g. Fear of reprisal. For centuries the workers have been told what to do by their superiors. The Just-In-Time process requires the workers to make many of the same decisions the managers would have made. The first time a mistake is made, the manager must not censure the workers, which may cause them to retract from the whole process.

9. Keep the Momentum Going

Several things can be done to keep the momentum going during and after the early phases of the Just-In-Time implementation.

a. External pressure for improvement. This usually comes from the economic conditions of the marketplace or the competitors.

b. Top management commitment. This is one of the greatest failings of poor Just-In-Time implementations.

c. Use of change agents. Often corporate or divisional staff, consultants or trainers can be used to facilitate the conversion to Just-In-Time.

d. Organizationwide involvement. If the workers see that the Just-In-Time is all around them, the peer pressure and their desire to keep up with their job requirements will motivate them to work within the plan.

e. Experimentation and innovation. As new ideas are generated, many through trial and error, especially where the workers have had an input and participation in the testing and research, they will be easier to implement and facilitate greater ownership by the worker in the new process.

f. Reinforcement—early and often—must occur. The minute criticism becomes apparent from leaders, the sooner the Just-In-Time project will die.

10. Some Examples from a Case Study

At one computer manufacturer that was implementing Just-In-Time, the simplicity of the JIT concepts seemed to be understood; but the people had considerable difficulty in changing their habits, as well as changing the processes continuously. Human nature resists change, especially when one has just completed the last change and has begun to see results. As these constant changes occurred, the workers in this plant became resistant. Considerable resistance developed when layoffs occurred because of economic reasons, when management changed the definitions for quality, and when external support groups failed to work with the changes.

It was discovered that most attendees were frustrated and confused about the JIT *applications—not* the concepts. They saw inconsistent management support, arbitrary demand changes, inaccurate planning data, excessive demands on their skills, little positive feedback, and poor recognition for what they had already accomplished. It was also found, through testing and using *the Just-In-Time Self Test*, that the middle-level production supervisors were the most resistant. Upper-level management and the workers were fairly well motivated. So some changes were facilitated in supervisory positions and progress was resumed; it was called JIT "career counseling."

The phrase "Success comes in small lots" characterized much of the problem solving that occurred in their implementation. The elements that were focused on were setup reduction, forming product family group cells, balancing assembly cycle times to other flows, planning for excess capacity and worker cross training. These are typical activities that occur during any implementation.

At this company, *it would never have happened without the Champions!*—especially the production manager and the quality assurance manager. A "JIT Council" was established, composed of most upper-level managers and directors to decide on policy. The two champions at this level were the single biggest reason JIT was implemented. Even at that, they weren't very effective until after the training of the employees and the collection of *their* input about the flow problems.

One would assume these improvements would essentially guarantee permanent survival of this company in the marketplace. Survival is one of the great motivators of JIT implementations, and it was certainly the motivator in this company. However, history caught up with them. The company was acquired by a large international firm, which subsequently made a decision to shut down this division and disband it. It had no desire to be in the mini-computer market, no matter how well the division had performed. However, the employees were given several months to find other employment. They were snapped up immediately. Their tested and proven experience at Just-In-Time manufacturing was a valuable asset to every one of them and made them highly desirable to the rest of the industry.

Now you've done it. You've been through the whole book. Please don't let all this knowledge wander off without some action on your part to improve your environment. Everyone can benefit from your initiative. Good luck on your Just-In-Time journey! Fill out the Just-In-Time Project Plan and get started!!

Just-In-Time Project Plan

List items you discovered in Chapters Five through Twelve:

Education and People
1. _____
2. _____
3. _____
4. _____
5. _____
6. _____
7. _____
8. _____
9. _____
10. _____

Quality Management
1. _____
2. _____
3. _____
4. _____
5. _____
6. _____
7. _____
8. _____
9. _____
10. _____

Chapter 13/Building an Implementation Plan

Factory Flow
1. _____
2. _____
3. _____
4. _____
5. _____
6. _____
7. _____
8. _____
9. _____
10. _____

Production Processes
1. _____
2. _____
3. _____
4. _____
5. _____
6. _____
7. _____
8. _____
9. _____
10. _____

Master Planning
1. _____
2. _____
3. _____
4. _____
5. _____
6. _____
7. _____
8. _____
9. _____
10. _____

Purchasing

1. _____
2. _____
3. _____
4. _____
5. _____
6. _____
7. _____
8. _____
9. _____
10. _____

Data Integrity

1. _____
2. _____
3. _____
4. _____
5. _____
6. _____
7. _____
8. _____
9. _____
10. _____

Results

1. _____
2. _____
3. _____
4. _____
5. _____
6. _____
7. _____
8. _____
9. _____
10. _____

Glossary

aggregate inventory The inventory for any groupings of items or products involving multiple stockkeeping units.

algorithm A prescribed set of well-defined rules or processes for the solution of a problem in a finite number.

allocation In an MRP system, an allocated item is one for which a picking order has been released to the stockroom but not yet sent out of the stockroom. It is an uncashed stockroom requisition. It can also be a process used to distribute material in short supply.

alpha factor The smoothing constant applied to the most recent forecast error in exponential smoothing forecasting.

anticipation inventories Additional inventory above basic stocking levels to cover projected increasing trends of sales, planned sales promotion programs, seasonal fluctuations, plant shutdowns, price changes, shortages and vacations.

assemble-to-order product A make-to-order product where all components (bulk semifinished, intermediate, subassembly, fabricated, purchased, packaging, etc.) used in the assembly, packaging, or finishing process are planned and stocked in anticipation of a customer order. *See* make-to-order product.

assembly A group of subassemblies and/or parts that are put together and constitute a major subdivision for the final product. An assembly may be an end item or a component of a higher-level assembly. *See* component; subassembly.

automated storage/retrieval system A high-density rack storage system with vehicles automatically loading and unloading the racks.

available inventory The onhand balance minus allocations, reservations, backorders, and (usually) quantities held for quality problems. Often called "beginning available balance."

available to promise The uncommitted portion of a company's inventory or planned production. This figure is normally calculated from the master production schedule plus the beginning balance in the first planning period less all customer orders before the next master scheduled receipt. It is used as a tool for customer order promising in assemble-to-order and make-to-order operations.

average forecast error The arithmetic mean of the forecast errors, or the exponentially smoothed forecast error. *See* forecast error; mean absolute deviation.

backflush The deduction from inventory of the component parts used in an assembly or subassembly by exploding the bill of materials by the production count of products or assemblies produced. *Syn:* postdeduct inventory transaction processing or theoretical consumption. *See* single-level backflush; superflush.

backlog All of the customer orders received but not yet shipped. Sometimes referred to as "open orders" or the "order board."

backorder An unfilled customer order or commitment. It is an immediate (or past due) demand against an item whose inventory is insufficient to satisfy the demand. *See* stockout.

back scheduling A technique for calculating operation start and due dates. The schedule is computed starting with the due date for the order and working backward to determine the required start date and/or due dates for each operation.

balancing operations In repetitive Just-In-Time production, trying to match actual output cycle times of all operations to the demand of use for parts as required by final assembly and eventually as required by the market.

bar coding A method of encoding data for fast and accurate readability. Bar codes are a series of wide or narrow, light or dark, vertically printed lines where the width of lines and spaces between lines is arranged to represent letters and numbers.

beginning inventory A statement of the inventory count at the end of last period, usually from a perpetual inventory record.

bill of labor *See* product load profile.

bill of material (BOM) A listing of all the subassemblies, intermediates, parts, and raw materials that go into a parent assembly showing the quantity of each required to make an assembly. There are a variety of display formats of bill of material, including single level bill of material, indented bill of material, modular (planning) bill of material, transient bill of material, matrix bill of material, costed bill of material, etc. May also be called "formula," "recipe," "ingredients list" in certain industries.

bill of material structuring The process of organizing bills of material to perform specific functions. *See* bill of material, planning bill.

bill of resources *See* product load profile.

blanket order A long-term commitment to a vendor for material against which short-term releases will be generated to satisfy requirements. Often blanket orders cover only one item, with predetermined delivery dates.

blow-through *See* phantom bill of material.

bottleneck A facility, function, department, and so on that impedes production—for example, a machine or work center where jobs arrive at a faster rate than they can be completed.

bucketless system An MRP, DRP, or other time-phased system in which all time-changed dates are processed, stored, and usually displayed using dated records rather than defined time periods or "buckets." *See* bucketed system, time bucket.

burden Costs incurred in the operation of a business that *can* be directly related to the individual work centers for products or services produced. These costs, such as power consumption, indirect labor, engineering support, and set-up people, are charged or allocated directly to the work center that employs them. These costs are then distributed to units of product, or service, by allocating the proportion of costs that were used by any product or service flowing through that work center. This is the basis of activity based costing.

capacity (1) In a general sense, refers to an aggregated volume of work load. It is a separate concept from priority. *See* priority. (2) The highest reasonable output rate that can be achieved with the current product specifications, product mix, work force, plant, and equipment.

capacity requirement planning (CRP) The function of establishing, measuring, and adjusting limits or levels of capacity. The term *capacity requirements planning* in this context is the process of determining how much labor and machine resources are required to accomplish the tasks of production. Open shop orders, and planned orders in the MRP system, are input to CRP, which "translates" these orders into hours of work by work center by time period. *See* infinite loading, closed-loop MRP; rough cut capacity planning.

carrying cost Cost of carrying inventory, usually defined as a percent of the dollar value of inventory per unit of time (generally one year). Depends mainly on cost of capital invested as well as the costs of maintaining the inventory, such as taxes and insurance, obsolescence, spoilage, and space occupied. Such costs vary from 10–35 percent annually, depending on type of industry. Ultimately, carrying cost is a policy variable reflecting the opportunity cost alternative uses for funds tied up in inventory. *See* economic order quantity; cost of capital.

Cause and Effect Diagram A precise statement of a problem or phenomenon with a branching diagram leading from the statement to the known potential causes. *Syn:* fishbone chart, Ishikawa diagram.

Cellular Manufacturing A manufacturing process that produces families of parts within a single line or cell of machines operated by machinists who work only within the line or cell.

centralized dispatching Organization of the dispatching function into one central location. This often involves the use of data collection devices for communication between the centralized dispatching function, which usually reports to the production control department, and the shop manufacturing departments. *See* decentralized dispatching.

changeover The refitting of equipment to either neutralize the effects of the just completed production, to prepare the equipment for production of the next scheduled item, or both. *See* setup.

changeover cost The sum of the teardown costs and the setup costs for a manufacturing operation. *See* idle time. *Syn:* turnaround costs; shutdown/startup costs.

closed-loop MRP A system built around material requirement planning and also including the additional planning functions of sales and operations (production planning, master production scheduling, and capacity requirements planning). Further, once this planning phase is complete and the plans have been accepted as realistic and attainable, the execution functions come into play. These include the manufacturing control functions of input-output measurement, detailed scheduling and dispatching, as well as anticipated delay reports from both the plant and vendors, vendor scheduling, etc. The term *closed-loop* implies that not only is each of these elements included in the overall system but also that there is feedback from the execution functions so the planning can be kept valid at all times. *See:* manufacturing resource planning.

commodity buying Grouping, like parts or materials, under one buyer's control for the procurement of all requirements to support production.

component A term used to identify a raw material, ingredient, part, or subassembly that goes into a high-level assembly, compound, or other item. May also include packaging materials for finished items. *See* assembly.

composite part A part that represents operations common to a family or group of parts controlled by group technology. Tools, jigs, and dies are used for the composite part and, therefore, any parts of that family can be processed with the same operations and tooling. The goal here is to reduce setup costs. *See* group technology.

computer aided design (CAD) The use of computers in interactive engineering drawing and storage of designs. Programs complete that layout, geometric transformations, projections, rotations, magnifications, and interval (cross-section) views of a part and its relationship with other parts.

computer aided manufacturing (CAM) Use of computers to program, direct, and control production equipment in the fabrication of manufactured items.

computer numerical control (CNC) A technique in which a machine tool control uses a minicomputer to store numerical instructions.

consigned stocks Inventories, generally of finished products, that are in the possession of customers, dealers, agents, etc., but remain the property of the manufacturer by agreement with those in possession.

constraint A limitation placed on the maximization or minimization of a objective function. These usually result from scarcity of the resources necessary for attaining some objective. *See* objective function. *Syn:* structural constraint; restriction.

continuous production A production system in which the productive equipment is organized and sequenced according to the steps involved to produce the product. Denotes that material flow is continuous during the production process. The routing of the jobs is fixed and setups are seldom changed. *See* intermittent production.

control chart A statistical device usually used for the study and control of repetitive processes. It is designed to reveal the randomness or non-randomness of deviations from a mean or control value, usually by plotting these.

cost center The smallest segment of an organization for which costs are collected, typically a department. The criteria in defining cost centers are that the cost be significant and the area of responsibility be clearly defined. A cost center may not be identical to a work center. Normally, it would encompass more than one work center. *See* work center.

cost of capital The cost of maintaining a dollar of capital invested for a certain period, normally one year. This cost is normally expressed as a percentage and may be based on such factors as the average expected return on alternative investments and current bank interest rate for borrowing. *See* economic order quantity.

critical path method (CPM) A network planning technique used for planning and controlling the activities in a project. By showing each of these activities and their associated times, the "critical path" can be determined. The critical path identifies elements that actually constrain the total time for the project. *See* PERT.

cumulative lead time The longest planned length of time involved to accomplish the activity in question. For any item planned through MRP, it is found by reviewing the lead time for each bill of material path below the item. Whichever path adds up to the greatest number defines cumulative lead time. *Syn:* aggregate lead time; stacked lead time; composite lead time; critical path lead time; and combined lead time.

cycle (1) The interval of time during which a system or process, such as seasonal demand or a manufacturing operation, periodically returns to similar initial conditions. In inventory control, a cycle is often taken to be the length of time between two replenishment shipments. (2) The interval of time during which an event or set of events is completed. In production control, a cycle is often taken to be the length of time between the release of a manufacturing order and shipment to the customer or inventory. *Syn:* manufacturing cycle; manufacturing lead time.

cycle counting An inventory accuracy audit technique where inventory is counted on a cyclic schedule, rather than once a year. For example, a cycle inventory count is usually taken on a regular, defined basis (often more frequently for high-value fast-moving items and less frequently for low-value or slow-moving items). Most effective cycle counting systems require the counting of a certain number of items every workday with each item counted at a prescribed frequency. The key purpose of cycle counting is to identify items in error, thus triggering research, identification, and elimination of the cause of the errors. *See* ABC classification.

cycle stock One of the four main components of any item inventory, the cycle stock is the most active part; (i.e., that which depletes gradually and is replenished cyclically when orders are received). *See* lot size.

cycle time (1) In industrial engineering, the time between completion of two discrete units of production. For example, the cycle time of motors assembled at a rate of 120 per hour would be 30 seconds, or one every half minute. (2) In materials management, it refers to the length of time from when material enters a production facility until it exists. *Syn:* throughput time.

data Any representations, such as alphabetic or numeric characters, to which meaning can be assigned. *See* information.

data base A data processing file-management approach designed to establish the independence of computer programs from data files. Redundancy is minimized and data elements can be added to or deleted from the file designs without necessitating changes to existing computer programs.

dedicated line A production line "permanently" configured to run well-defined parts, one piece at a time from station to station.

delivery lead time The time from the receipt of the customer order to the delivery of the product. *Syn:* delivery cycle. *See* lead time.

delivery schedule The required and/or agreed time or rate of delivery of goods or services purchased for a future period.

demand A need for a particular product or component. The demand could come from any number of sources, (i.e., customer order, forecast, interplant, branch warehouse, service part, or for manufacturing another product). At the finished goods level, "demand data" are usually different from "sales data" because demand does not necessarily result in sales, (i.e., if there is no stock there will be no sale). *See:* dependent demand; independent demand.

demand filter A standard set to monitor individual sales data in forecasting models. Usually set to be tripped when the demand for a period differs from the forecast by more than some number of mean absolute deviations.

demand management The function of recognizing and managing all of the demands for products to ensure that the master scheduler is aware of them. It encompasses the activities of forecasting, order entry, order promising, branch warehouse requirements, interplant orders, and service parts requirements. *See* master production schedule.

demand pull The triggering of material movement to a work center only when that work center is out of work and/or ready to begin the next job. It in effect eliminates queues from a front of a work center, but it can cause queues at the end of a previous work center. *See* pull.

demonstrated capacity Proven capacity calculated from actual output performance data, usually number of items produced times standard hours per item. *Syn:* actual capacity. *See* rated capacity.

dependent demand Demand is considered dependent when it is directly related to or derived from the schedule for other items or end products. Such demands, therefore, are calculated and need not and should not be forecast. A given inventory item may have both dependent and independent demand at any given time. *See* independent demand.

deterioration Product spoilage, damage to the package, and the like. One of the considerations in inventory carrycost. *See* obsolescence.

deviation The difference, usually the absolute difference, between a number and the mean of a set of numbers, or between a forecast value and the actual datum.

direct cost A variable cost that can be directly attributed to a particular job or operation. *See:* variable cost.

direct labor Labor specifically applied to the product being manufactured or utilized in the performance of the service.

direct material Materials which become a part of the final product in measurable quantities. *See* indirect materials.

discrete manufacturing Production of distinct items: automobiles, appliances, computers. *See* process manufacturing.

distribution requirements planning The function of determining the needs to replenish inventory at branch warehouses. A time-phased order point approach is used, where the planned orders at the branch warehouse level are "exploded" via MRP logic to become gross requirements on the supplying source. In the case of multilevel distribution networks, this explosion process can continue down through the various levels of master warehouse, factory warehouse, and so on, and become input to the master production schedule. Demand on the supplying source(s) is recognized as dependent, and standard MRP logic applies. *See* time-phased order point; physical distribution; push (distribution system).

downstream operation Subsequent tasks to the task currently being planned or executed.

downtime Time when a machine is scheduled for operation but is not producing for such reasons as maintenance, repair, or setup.

due date The date when purchased material or production material is due to be available for use. *See* need date; scheduled receipt.

dynamic lot sizing A lot-sizing technique that creates an order quantity subject to continuous recomputation. *See* least total cost; least unit cost; part period balancing; period order quantity.

Economic Order Quantity (EOQ) A type of fixed order quantity, which determines the amount of an item to be purchased or manufactured at one time. The intent is to minimize the combined costs of acquiring and carrying inventory.

efficiency Standard hours earned divided by actual hours worked. Efficiency is a measure of how closely predetermined standards are achieved. Efficiency for a given time can be calculated for a machine, an employee, a group of machines, a department, and so on.

electronic data interchange (EDI) The paperless (electronic) exchange of trading documents such as purchase orders, shipment authorizations, advance shipment notices, invoices, or sales orders using standardization document formats. EDI is used mostly between businesses and institutions with different computer systems.

ending inventory A statement of onhand quantities at the end of a period, often determined by a physical inventory. *See* beginning inventory.

engineering change A revision to a parts list, bill of material, or drawings authorized by the engineering department. Changes are usually identified by a control number and are made for safety, cost reduction, functionality, or process improvement reasons. To effectively implement engineering changes, all affected functions, such as materials, purchasing, quality assurance, and assembly engineering, should review and agree to the changes.

engineer-to-order Products whose customer specifications require unique engineering design or significant customization. Each customer order then results in a unique set of part numbers, bills of material, and routings.

exception report A report that lists or flags only those items deviating from plan.

expediting The "rushing" or "chasing" of production or purchase orders that are needed in less than the normal lead time. In program review and evaluation technique (PERT), expediting is to track and report performance of each step in the process to determine if the project is on track, compared to a "critical path" of the longest lead time item sequence. *See* dispatching.

exponential smoothing A type of weighted moving average forecasting technique in which past observations are geometrically discounted according to their age. The heaviest weight is assigned to the most recent data. The smoothing is termed *exponential* because data points are weighted in accordance with an exponential function of their age. The technique makes use of a smoothing constant to apply to the difference between the most recent forecast and the critical sales data, that avoids the necessity of carrying historical sales data. The approach can be used for data that exhibit no trend or seasonal patterns or for data with either (or both) trend and seasonality.

exposures The number of times per year that the system risks a stockout. This number of exposures is arrived at by dividing the lot size into the annual usage. Theoretically, the smaller the lot size the larger the number of exposures to a stockout.

external setup time Elements of a setup procedure performed while the process is in production; the machine is running. *See* internal setup time.

extrinsic forecast A forecast based on a correlated leading indicator or indices such as estimating furniture sales based on housing starts. Extrinsic forecasts tend to be more useful for large aggregations, such as total company sales or product groups, than for individual product sales. *Ant:* intrinsic forecasts.

fabrication A term used to distinguish manufacturing operations for components, as opposed to assembly operations.

failsafe work methods Methods of performing operations so incorrect actions cannot be completed. Examples: part without holes in the proper place cannot be removed from a jig; a three-pronged electrical plug that can only be plugged in one way. Called "poka-yoke" by the Japanese.

failure mode and effects analysis (FMEA) A methodology for analyzing products for failure types in order to prevent customer dissatisfaction. Potential failure modes or types are developed by process and design teams. The effects on customer and process are estimated. Causes are defined and occurence, detection capability, and severity are assessed. A risk priority is assigned and corrective action is instituted. An FMEA report format is used to facilitate this activity.

families A group of end items whose similarity of design and manufacture facilitates being planned in aggregate, whose sales performance is monitored together, and occasionally whose cost is aggregated at this level.

feedback The flow of information back into the control system so actual performance can be compared with planned performance.

file An organized collection of records or the storage device in which these records are kept.

final assembly (1) The highest level of assembled product, as it is shipped to customers. (2) The name for the manufacturing department where the product is assembled. *Syn:* erection department; blending department; packout department.

final assembly schedule (FAS) Also referred to as "finishing schedule," because it may include other operations than simply the final operations. It may also not involve "assembly" but only final mixing, cutting, packaging, and so on. It is a schedule of end items to finish the product for specific customer orders in a "make-to-order" or "assemble-to-order" environment. It is prepared after receipt of a customer order as constrained by the availability of material and capacity, and it schedules the operations required to complete the product from the level where it is stocked (or master scheduled) to the end-item level. *Syn:* blending; schedule; packout schedule.

firm planned order (FPO) A planned order that can be frozen in quantity and time. The computer is not allowed to automatically change it; this is the responsibility of the planner in charge of the item that is being planned. This technique can aid planners to respond to material, supplier, and capacity problems by firming selected planned orders. Additionally, firm planned orders are the normal method of stating the master production schedule.

fishbone chart *See* cause and effect diagram.

fitness for use Involves the quality of a product and/or the appropriateness of its design characteristics for its intended use by internal or external customers.

fixed cost An expenditure that does not vary with the production volume; for example: rent; property tax, salaries of certain personnel. *See* variable cost.

fixed order quantity A lot-sizing technique for inventory management that will always cause planned or actual orders to be generated for a predetermined fixed quantity or multiples thereof if net requirements for the period exceed the fixed order quantity. *See* economic order quantity; lot-for-lot; period order quantity.

flow control A term used to describe a specific production control system that is based primarily on setting production rates and feeding work into production to meet these planned rates, then following it through production to make sure that it is moving. Flow control has its most successful application in repetitive production. *See* order control.

flow shop A form of manufacturing organization in which machines and operators handle a standard, usually uninterrupted, material flow. The operators tend to perform the same operations for each production run. A flow shop is often referred to as a mass production shop, or is said to have a continuous manufacturing layout. The plant layout (arrangement of machines, benches, assembly lines, etc.) is designed to facilitate a product "flow." The process industries (chemicals, oil, paint, etc.) are extreme examples of flow shops. Each product, though variable in material specifications, uses the same flow pattern through the shop. Production is set at a given rate, and the products are generally manufactured in bulk.

fluctuation inventory Inventories that are carried as a cushion to protect against forecast error. *See* safety stock.

focus forecasting A system that allows the user to simulate the effectiveness of numerous forecasting techniques, thereby being able to select the most effective one.

focused factory A plant or production group set up as a separate entity to produce a specific number of products and/or employ a limited number of processes.

forecast An estimate of future demand. A forecast can be determined by mathematical means using historical data; it can be created subjectively by using estimates from informal sources; or it can represent a combination of both techniques. *See* extrinsic forecast; intrinsic forecast.

forecast consumption The process of replacing the forecast with customer orders, or other types of actual demands, as they are received.

forecast error The difference between actual demand and forecast demand, stated as an absolute value or as a percentage.

gantt chart A control chart especially designed to show graphically the relationship between planned performance and actual performance, named after its originator, Henry L. Gantt. Used for machine loading, where one horizontal line is used to represent capacity and another to represent load against that capacity, or for following job progress where one horizontal line represents the production schedule and another parallel line represents the actual progress of the job against the schedule in time. Often used in project planning. *Syn:* job progress chart.

gross requirement The total of independent and dependent demand for a component prior to the netting of onhand inventory and scheduled receipts.

group technology An engineering and manufacturing philosophy that identifies the "sameness" of parts, equipment, or processes. It provides for rapid retrieval of existing designs and anticipates a cellular type production equipment layout.

handling cost The cost involved in handling inventory. In some cases, the handling cost incurred may depend on the size of the inventory.

heuristic A form of problem solving where the results or rules have been determined by experience or intuition instead of by optimization.

idle time Time when operators or machines are not producing product because of setup, maintenance, lack of material, tooling. *Syn:* downtime.

inbound stockpoint A defined location next to the place of use on a production flow to which materials are brought as needed, and from which material is taken for immediate use. Used with a pull system of material control. *See* Just-in-Time.

indented bill of material A form of multilevel bill of material. It exhibits the highest-level parents closest to the left side margin, and all the components going into these parents are shown indented to the right of the margin. All subsequent levels of components are indented farther to the right. If a component is used in more than one parent within a given product structure, it will appear more than once, under every subassembly in which it is used.

independent demand Demand for an item is considered independent when such demand is unrelated to the demand for other items. Demand for finished goods, parts required for destructive testing, and service parts requirements are some examples of independent demand. *See* dependent demand.

independent demand inventory system The policies, methods, and procedures used to manage inventory items that have independent demand. *See* independent demand; order point system.

indirect cost Cost not directly incurred by a particular job or operation. Certain utility costs, such as plant heating, are often indirect. An indirect cost is typically distributed to the product through the overhead rates. *See* direct cost.

indirect labor Work required to support production in general without being related to a specific product; for example, floor sweeping.

indirect materials Materials that become part of the final product but in such small quantities their cost is not applied directly to the product. Instead, their expense becomes a part of manufacturing supply or overhead costs.

infinite loading Showing the work behind work centers in the time periods required regardless of the capacity available to perform this work. The term *infinite loading* is considered to be obsolete today, although the specific computer programs used to do infinite loading can now be used to perform the technique called "capacity requirements planning." Infinite loading was a gross misnomer to start with, implying that a load could be put into a factory regardless of its availability to perform. The poor terminology obscured the fact that it is necessary to generate capacity requirements and compare these with available capacity before trying to adjust requirements to capacity. *See* capacity requirements planning; finite loading.

internal setup time Elements of a setup procedure performed while the process is not running. *Ant:* external setup time.

interoperation time The time between the completion of one operation and the start of the next. *See* move time.

interplant demand Items to be shipped to another plant or division within the corporation. Although it is not a customer order, it is usually handled by the master production scheduling system in a similar manner. *See* demand management.

intransit inventory Material moving between two or more locations, usually separated geographically; for example, the shipment of finished goods from a plant to a distribution center.

inventory Items that are in a stocking location or work in process and that serve to decouple successive operations in the process of manufacturing a product and distributing it to the consumer. Inventories may consist of finished goods ready for sale; they may be parts or intermediate items; they may be work in process; or they may be raw materials.

inventory buffer *See* fluctuation inventory.

inventory control The activities and techniques of maintaining the stock of items at desired levels, whether they are raw materials, work in process, or finished products.

inventory investment The number of dollars that are tied up in all levels of inventory.

inventory turnover The number of times that an inventory "turns over," or cycles during the year. A frequently used method to compute inventory turnover is to divide the average inventory level into the annual cost of sales; for example, if average inventory were $3 million and cost of sales were $21 million, the inventory would be considered to "turn" seven times per year.

inventory valuation The value of the inventory at either its cost or its market value. Because inventory value can change with time, some recognition is taken of the age distribution of inventory. Therefore, the cost of value of inventory is usually computed on a first-in-first-out (FIFO), or a last-in-first-out (LIFO) basis, or a standard cost basis to establish the cost of goods sold.

Ishikawa diagram *See* cause-and-effect diagram.

issue The physical movement of items from a stocking location. Often, also refers to the transaction reporting of this activity. *See* planned issue.

item Any unique manufactured or purchased part, material, intermediate, subassembly, or product.

item number A number that serves to uniquely identify an item. *Syn*: part number; stock code.

item record The "master" record for an item. Typically it contains identifying and descriptive data, control values (lead times, lot sizes) and may contain data on inventory status, requirements, planned orders, and costs. Item records are linked together by bill of material records (or product structure records), thus defining the bill of material.

jidoka Practice of stopping the production line when a defect occurs.

JIT *See* Just-in-Time.

job order costing A costing system in which costs are collected to specific jobs. Also called "job costing." This system can be used with either actual or standard costs in the manufacturing of distinguishable units or lots of products.

job shop Typically a make-to-order manufacturning plant. Products are made in intermittent sequences as defined by specific customer orders. *Ant:* flow shop. *See* make-to-order product.

Just-in-Time (JIT) Having only the correct quality part in the right place at the right time. The implication is that each operation is so closely synchronized with the subsequent ones that the signal to perform work in any station is from the downstream station. The only inventory in the work being performed not between stations. In the broad sense, an approach to achieving excellence in a manufacturing company based on the continuing elimination of waste (waste being considered as those things that do not add value to the product). *See* zero inventories.

kanban A method of signaling in Just-in-Time production that uses standard containers or lot sizes with a single card attached to each. It is a pull system in which work centers signal with a card that they wish to withdraw parts from feeding operations or suppliers. Kanban, in Japanese, loosely translated means "card," literally "billboard" or "sign." The term is often used synonymously for the specific scheduling system developed and used by Toyota Corporation in Japan.

kit The components of a parent that have been pulled from stock and readied for movement to a production area.

lead time A span of time required to perform an activity. In a logistics context, the time between recognition of the need for an order and the receipt of goods. Individual components of lead time can include: order preparation time, queue time, move or transportation time, receiving and inspection time. *See* manufacturing lead time; purchasing lead time.

lead time offset A technique used in MRP where a planned order receipt in one time period will require the release of that order in an earlier time period based on the lead time for the item.

learning curve A planning technique particularly useful in the project-oriented industries where new products are phased in rather frequently. The basis for the learning curve calculation is the fact that workers will be able to produce the product more quickly after they get used to making it. For example, in a 90 percent learning curve, every time the quantity to make doubles, the time to do the work drops to 90 percent of the previous time to do the work.

level Every part or assembly in a product structure is assigned a level code signifying the relative level in which that part or assembly is used within that product structure. Normally, the end items are assigned level "0" with the components/subassemblies going into it assigned level "1" and so on. MRP explosion process starts from level "0" and proceeds downward one level at a time.

line balancing The assignment, organizing, and equalizing of sequential tasks through workstations to minimize the number of workstations and idle time at all stations, to balance (match) the amount of work time at each station. Any assembly line process can be divided into elemental tasks, each with a specified time requirement per unit of product and a sequence relationship with the other tasks. Line balancing can also mean a technique for determining the product mix that can be run down an assembly line to provide a fairly consistent flow of work through that assembly line at the planned line rate.

load leveling Spreading orders out in time or so rescheduling operations that the amount of work to be done in sequential time periods tends to be distributed evenly and is achievable. *See* finite loading.

logistics In an industrial context, this term refers to the art and science of obtaining and distributing material and product. In a military sense, its meaning can also include the movement of personnel and spare parts.

lot A quantity produced together and sharing the same production costs and resultant specifications.

lot-for-lot A lot-sizing technique that generates planned orders in quantities equal to the net requirements in each period. *Syn:* discrete order quantity.

lot number A unique identification assigned to a homogenous quantity of material. *Syn:* batch number; mix number.

lot number control Assignment of unique numbers to each instance of receipt and carrying forth that number into subsequent manufacturing processes so that, in review of an end item, each lot consumed from raw materials through end item can be identified as having been used for the manufacture of this specific end item lot.

lot size The amount of a particular item that is ordered from the plant or a supplier. *Syn:* order quantity.

lot traceability The ability to identify the lot or batch numbers of consumption and/or composition for manufactured, purchased, and shipped items. This is a federal requirement in certain regulated industries.

low level code Identifies the lowest level in any bill of material at which a particular component may appear. Net requirements for a given component are not calculated until all the gross requirements have been calculated down to that level. Low level codes are normally calculated and maintained automatically by the computer software. *See* level.

lumpy demand A demand pattern with large fluctuations from one time period to another. *Syn:* discontinuous demand.

maintenance repair and operating supplies (MRO) Items used in support of general operations and maintenance, such as maintenance supplies, spare parts, consumables used in the manufacturing process.

major setup The equipment setup and related activities required to manufacture a group of items in sequence, exclusive of the setup required for each item in the group. *Ant:* minor setup.

make-or-buy decision The act of deciding whether to produce an item in-house or buy it from an outside supplier.

make-to-order product A product that is finished after receipt of a customer order. Frequently long lead time components are planned prior to the order arriving to reduce the delivery time to the customer. Where options or other subassemblies are stocked prior to customer orders arriving, the term *assemble-to-order* is frequently used. *See* job shop.

make-to-stock product A product shipped from finished goods, "off the shelf," and therefore is finished prior to a customer order arriving.

manufacturing calendar A calendar, used in inventory and production planning functions, that consecutively numbers only the working days so the component and work order scheduling may be done based on the actual number of work days available. *Syn:* M-day calendar.

manufacturing lead time The total time required to manufacture an item, exclusive of purchasing lead time. Included here are order preparation time, queue time, setup time, run time, move time, inspection, and put-away time.

manufacturing order A document, group of documents, or schedule identity conveying authority for the manufacture of specified parts or products in specified quantities.

manufacturing process The series of activities performed on a material to convert it from the raw or semifinished state to a state of further completion and a greater value.

manufacturing resource planning (MRP II) A method for the effective planning of all resources of a manufacturing company. Ideally, it addresses operational planning in units, financial planning in dollars, and has a simulation capability to answer "what if" questions. It is made up of a variety of functions, each linked together: business planning, sales and operations (production planning), master production scheduling, material requirements planning, capacity requirements planning, and the execution support systems for capacity and material. Output from these systems would be integrated with financial reports, such as the business plan, purchase commitment report, shipping budget, inventory projection in dollars, and so on. Manufacturing resource planning is a direct outgrowth and extension of closed-loop MRP. *See* closed-loop MRP; material requirements planning.

manufacturing strategy A collective pattern of decisions that act on the formulation and deployment of manufacturing resources. To be most effective, the manufacturing strategy should act in support of the overall strategic direction of the business unit and provide competitive advantage where called for.

master production schedule (MPS) The anticipated build schedule for those items assigned to the master scheduler. The master scheduler maintains this schedule and, in turn, it becomes a set of planning numbers that "drives" material requirements planning. It represents what the company plans to produce expressed in specific configurations, quantities, and dates. The master production schedule is not a sales forecast, which represents a statement of demand. The master production schedule must take into account the forecast, the production plan, and other important considerations, such as backlog, availability of material, availability of capacity, management policy, and goals. *Syn:* master schedule. *See* closed-loop MRP.

master scheduler The job title of the person who manages the master production schedule. This person should be the best scheduler available, because the consequences of the planning done here have a great impact on material and capacity planning. Ideally, the person would have substantial product and plant knowledge.

material planner The person normally responsible for managing the inventory levels, schedules, and availability of selected items, either manufactured or purchased. In an MRP system, the person responsible for reviewing and acting on order release, action, and exception messages from the system.

material requirements planning (MRP) A set of techniques that uses bills of material, inventory data, and the master production schedule to calculate requirements for materials. It makes recommendations to release replenishment orders for material. Further, since it is time phased, it makes recommendations to reschedule open orders when due dates and need dates are not in phase. Originally seen as merely a better way to order inventory, today it is thought of as primarily a scheduling technique, (i.e., a method for establishing and maintaining valid due dates—priorities) on orders. *See* closed-loop MRP; manufacturing resource planning.

materials management The grouping of management functions supporting the complete cycle of material flow, from the purchase and internal control of production materials to the planning and control of work in process to the warehousing, shipping, and distribution of the finished product.

material usage variance The difference between the planned or standard requirements for materials to produce a given item and the actual quantity used for this particular instance of manufacture.

material yield The ratio of usable material from a given quantity of same. *Syn:* potency.

matrix bill of material A chart made up from the bills of material for a number of products in the same or similar families. It is arranged in a matrix with components in columns and parents in rows (or vice versa) so requirements for common components can be summarized conveniently.

mean The arithmetic average of a group of values.

mean absolute deviation (MAD) The average of the absolute values of the deviations of some observed value from some expected value. MAD can be calculated based on observations and the arithmetic mean of those observations. An alternative is to calculate absolute deviations of actual sales data minus forecast data. These data can be averaged in the usual arithmetic way or with exponential smoothing.

min-max system A type of order point replenishment system where the "min" is the order point and the "max" is the "order-up-to" inventory level. The order quantity is variable and is the result of the "max" minus onhand and onorder inventory. An order is recommended when the onhand and onorder inventory is at or below the "min."

mixed-model production Making several different parts or products in varying lot sizes so a factory is making close to the same mix of products that will be sold that day. The mixed-model schedule governs the making and the delivery of component parts, including outside suppliers. The goal is to build every model, every day, according to demand.

modular bill (of material) A type of planning bill that is arranged in product modules or options. Often used in companies where the product has many optional features, (e.g., automobiles). *See* planning bill;

common parts bill; super bill.

move card In a Just-in-Time context, refers to a card or other signal indicating that a specific number of units of a particular item are to be taken from a source (usually outbound stockpoint) and taken to a point of use (usually inbound stockpoint). *Syn:* move signal. *See* kanban; production card.

move order The authorization to move a particular item from one location to another.

move ticket A document used in dispatching to authorize and/or record movement of a job from one work center to another. It may also be used to report other information, such as the active quantity or the material storage location.

MPS *See* master production schedule.

MRP *See* material requirements planning.

MRP II *See* manufacturing resource planning.

multilevel bill of material A display of all the components directly or indirectly used in a parent, together with the quantity required of every component. If a component is a subassembly, blend, intermediate, and the like, all of its components will also be exhibited and all of their components, down to purchased parts and materials.

multilevel where used A display for a component listing all the parents in which that component is directly used and the next higher level parents into which each of those parents is used, until ultimately all top-level (level 0) parents are listed.

need date The date when an item is required for its intended use. In an MRP system, this date is calculated by a bill of material explosion of a schedule and the netting of available inventory against that requirement. *See* due date.

net requirements In MRP, the net requirements for a part or an assembly are derived as a result of applying gross requirements and allocations against inventory on hand, scheduled receipts, and safety stock. Net requirements, lot sized and offset for lead time, become planned orders.

netting The process of calculating net requirements.

normal distribution A particular statistical distribution where most of the observations fall fairly close to one mean, and a deviation from the mean is as likely to be plus as it is likely to be minus. When graphed, the normal distribution takes the form of a bell-shaped curve.

obsolescence Loss of product value resulting from a model or style change or technological development. *See* deterioration.

onhand balance The quantity shown in the inventory records as being physically in stock. *See* available inventory.

onorder The quantity of stock onorder represented by the total of all outstanding replenishment orders. The onorder balance increases when a new order is released and it decreases when material is received against an order, or when an order is cancelled. *See* onhand balance; open order.

open order (1) A released manufacturing order or purchase order. *Syn:* scheduled receipt. (2) An unfilled customer order.

operation duration The total time that elapses between the start of setup of an operation and the completion of the operation.

operation priority The scheduled due date and/or start date of a specific operation for a specific job, usually as determined by the back scheduling process. *See* back scheduling; dispatching; order priority.

operation sequencing A simulation technique for short-term planning of actual jobs to be run in each work center based on capacity, priority, existing manpower, and machine availability.

operation sequence The sequential steps for an item to follow in its flow through the plant. For instance, operation 1: cut bar stock; operation 2: grind bar stock; operation 3: shape; operation 4: polish; operation 5: inspect and send to stock. This information is normally maintained in the routing file.

operating start date The date when an operation should be started for its operation and order due date to be met. It can be calculated based on scheduled quantities and lead times (queue, setup, run, move) or based on the work remaining and the time remaining to complete the job.

order entry The process of accepting and translating what a customer wants into terms used by the manufacturer or distributor. This can be as simple as creating shipping documents for a finished good product line, to a more complicated series of activities including engineering effort for make-to-order products.

ordering cost Used in calculating economic order quantities, and it refers to the costs that increase as the number of orders placed increases. Includes costs related to the clerical work of preparing, releasing, following, and receiving orders, the physical handling of goods, inspections, and setup costs, as applicable. *Syn:* acquisition cost. *Ant:* carrying costs.

order multiples An order quantity modifier applied after the lot size has been calculated that increments the order quantity to a predetermined multiple.

order point A set inventory level where, if the total stock onhand plus onorder falls to or below that point, action is taken to replenish the stock. The order point is normally calculated as historical or forecasted usage during the replenishment lead time plus safety stock. *Syn:* reorder point; trigger level. *See* time-phased order point.

order point system The method whereby replenishment orders are triggered (launched) at a calculated inventory level. *See* order point; time-phased order point; two bi; min-max.

order promising The process of making a delivery commitment (i.e., answering the question, "When can you ship?"). For make-to-order products, this usually involves a check of uncommitted material and availability of capacity. *Syn:* order dating; customer order promising. *See* available-to-promise.

output (1) Work being completed by a production facility. (2) The result of a computer program.

overhead Costs incurred in the operation of a business that cannot be directly related to the individual products or services produced. These costs, such as light, heat, supervision, and maintenance, are grouped in several pools (department overhead, factory overhead, general overhead) and distributed to units of product, or service, by some standard method, such as direct labor hours, direct labor dollars, direct materials dollars. *See* Burden.

paperless purchasing A purchasing operation that does not employ purchase requisitions or hard-copy purchase orders. In actual practice, a small amount of "paperwork" usually remains, normally in the form of the vendor schedule. *See* Just-in-Time; supplier scheduling.

Pareto's law A concept developed by Vilfredo Pareto, an Italian economist, which states that a small percentage of a group accounts for the largest fraction of the impact, value, and so on; for example, 20 percent of the inventory items may comprise 80 percent of the inventory value. *Syn:* 80/20 rule. *See* ABC classification.

part Normally refers to a material item used as a component and is not an assembly, subassembly blend, intermediate, and so on. *See* component.

part number *See* item number.

past due An order that has not been completed on the date scheduled. *Syn:* delinquent.

performance standard A criterion or benchmark against which actual performance is compared.

periodic inventory A physical inventory taken at some recurring interval (e.g., monthly, quarterly, or annual physical inventory).

perpetual inventory An inventory recordkeeping system where each transaction in and out is recorded and a new balance is computed.

perpetual inventory record A computer record or manual document on which each inventory transaction is posted so an accurate record of the inventory is maintained. *See* onhand balance.

phantom bill of material A bill of material coding and structuring technique used primarily for transient (nonstocked) subassemblies. For the transient item, lead time is set to zero and the order quantity to lot-for-lot. This permits MRP logic to drive requirements straight through the phantom item to its components, but usually retains its ability to net against any occasional inventories of the item. This technique also facilitates the use of common bills of material for engineering and manufacturing. *Syn:* transient bill of material; blow-though.

physical inventory (1) The actual inventory itself. (2) The determination of inventory quantity by actual count. Physical inventories can be taken on a continuous, periodic, or annual basis. *See* cycle counting.

picking The process of withdrawing from stock the components to make the products, or the finished goods to be shipped to a customer.

planned issue A disbursement of an item predicted by MRP through the creation of a gross requirement or allocation.

planned load The standard hours of work required by MRP recommended (planned) production orders. *See* planned order.

planned order A suggested order quantity, release date, and due date created by MRP processing, when it encounters net requirements. Planned orders are created by the computer, exist only within the computer, and may be changed or deleted by the computer during subsequent MRP processing if conditions change. Planned orders at one level will be exploded into gross requirements for components at the next lower level. Planned orders, along with released orders, also served as input to capacity requirements planning, to show the total capacity requirements in future time periods. *See* firm planned order; open order; scheduled receipt.

planned receipt A receipt against an open purchase order or open production order that has not yet been received.

planning bill (of material) An artificial grouping of items and/or events, in bill of material format, used to facilitate master scheduling and/or material planning. *See* common parts bill; modular bill; super bill.

planning horizon The span of time from the current to some future point for which plans are generated.

post-deduct *See* backflush

poka-yoke Mistake-proofing techniques, such as manufacturing or setup activity designed in a way to prevent an error from resulting in a product defect. For example, in an assembly operation, if each correct part is not used, a sensing device detects a part was unused and shuts down the operation, thereby preventing the assembler from moving the incomplete part on to the next station or beginning another one.

preventive maintenance A routine pattern of tests, checks, analysis, or replacements based on history, run time, cycles or operating characteristics. The purpose of these actions is to prevent unplanned shutdown. A good preventive maintenance program will extend equipment life, increase throughput, and increase productivity. Sometimes called Total Productive Maintenance (TPM).

primary work center The work center wherein an operation on a manufactured part is normally scheduled to be performed. *Ant:* alternate work center.

priority In a general sense, refers to the relative importance of jobs or activities (i.e., the sequence in time on which jobs should be worked). *See* capacity; scheduling.

priority control The process of communicating start and completion dates to manufacturing departments to execute a plan. The dispatch list is the tool normally used to provide these dates based on the current plan and status of all open orders.

priority planning The function of determining what material is needed and when. Master production scheduling and material requirements planning are the elements used for the planning and replanning process to maintain proper due dates on required materials.

process costing A cost accounting system in which the costs are collected by time and averaged over all the units produced during the period.

process flow production A production approach with minimal interruptions in actual processing in any one production run or between production runs of similar products. Queue time is virtually eliminated by integrating the moving of the product into the actual operation of the resource performing the work.

process manufacturing Production that adds value by mixing, separating, forming, and/or chemical reactions. It may be done in either batch or continuous mode.

process sheet Detailed manufacturing instructions issued to the plant. The instructions may include speeds, feed rates, temperatures, tools, fixtures, machines, inspections, and sketches of setups and semifinished dimensions. *See* routing.

process time The time during which the material is being worked, whether it is a machining operation or a hand assembly. *Syn:* residence time; cycle time.

procurement lead time The time required by the buyer to select a supplier and to place and obtain a commitment for specific quantities of material at specified times. The time between the receipt of a requirement to buy and the time the purchase is completed. *See* purchasing lead time.

product family A group of products with similar characteristics, often used in sales and operations (production) planning.

product group forecast A forecast for a number of similar products.

production card In a Just-in-Time context, it refers to a card or other signal for indicating that items should be made for use or to replace some items removed from pipeline stock. *See* kanban; move card.

production control The function of directing or regulating the movement of goods through the entire manufacturing cycle from the requisitioning of raw material to the delivery of the finished products. *See* inventory control.

production cycle The lead time to produce a product. *See* cycle.

production forecast A predicted level of customer demand for an option, feature, and so on, of an assemble-to-order product (or finish-to-order product). It is calculated by netting customer backlog against an overall family or product line master production schedule and then factoring this product "available-to-promise" by the option percentage in a planning bill of material.

production line A series of equipment or resources dedicated to the manufacture of a specific number of products or families.

production plan The agreed-on plan that comes from the sales and operation (production) planning function, specifically the overall level or rate of manufacturing output planned to be produced. Usually stated as a monthly rate for each product family (group of products, items, options, features, and so on). Various units of measure can be used to express the plan: units, tonnage, standard hours, number of workers. The production plan is management's authorization for the master scheduler to convert it into a more detailed plan; that is, the master production schedule. *See* sales and operations planning.

production schedule A plan that authorizes the factory to manufacture a certain quantity of a specific item. Usually initiated by the production planning department. *See* work order; manufacturing order.

productivity Refers to a relative measure of output per labor and/or machine input. An overall measure of production effectiveness is composed of two factors: efficiency (how well a resource is performing) and utilization (how intensively a resource is being used). It is calculated (1) as the product of the efficiency and utilization factors or (2) as the ratio of the output achieved as measured in standard hours to the total clock time scheduled for production for a given time period.

$$\text{Productivity} = \text{Efficiency} \times \text{Utilization}$$

or

$$\text{Productivity} = \text{Standard hours of output} \div \text{Clock time scheduled}$$

product layout *See* continuous production.

product load profile A listing of the required capacity and key resources required to manufacture one unit of a selected item or family. Often used to predict the impact of the item scheduled on the overall schedule and load of the key resources. Rough cut capacity planning uses these profiles to calculate the approximate capacity requirements of the master production schedule and/or the production plan. *Syn:* bill of labor; bill of resources; resource profile.

product mix The proportion of individual products that make up the total production and/or sales volume. Changes in the product mix can mean drastic changes in the manufacturing requirements for certain types of labor and material.

product structure The way components go into a product during its manufacture. A typical product structure or multilevel bill of material would show raw material converted into fabricated components, components put together to make subassemblies, subassemblies going into assemblies, and so forth.

projected available balance In MRP, the inventory balance projected into the future. It is the running sum of onhand inventory minus requirements plus scheduled receipts and planned orders. *See* projected on hand.

projected on hand Same as projected available balance, except excludes planned order.

pull (system) (1) In production, it refers to the production of items only as demanded for use, or to replace those taken for use. (2) In a material control context, it refers to the withdrawal of inventory as demanded by the using operations. Material is not issued until a signal comes from the user. (3) In distribution, it refers to a system for replenishing field warehouse inventories wherein replenishment decisions are made at the field warehouse itself, not at the central warehouse or plant. *Ant:* push system.

purchase part An item purchased from a supplier.

purchase requisition A document conveyed to the purchasing department to request the purchase of materials in specified quantities within a specified time.

purchasing capacity The act of buying capacity or machine time from a supplier. This allows a company to use and schedule the capacity of the machine or a part of the capacity of the machine as if it were in its own plant. This also shortens lead time by eliminating queue time at the supplier.

purchasing lead time The total lead time required to obtain a purchased item. Included here are order preparation and release time, supplier lead time, transportation time, receiving, inspections, and put-away time.

push (system) (1) In production, it refers to the production of items at times required by a given schedule planned in advance. (2) In material control, it refers to the issuing of material according to a given schedule and/or issued to a job order at its start time. (3) In distribution, it refers to a system for replenishing field warehouse inventories wherein replenishment decision making is centralized, usually at the manufacturing site or central supply facility. *Ant:* pull system.

quality Conformance to requirements.

quality at the source A producer's responsibility to provide 100 percent acceptable quality material to the consumer of the material. The objective is to reduce or eliminate shipping/receiving quality inspections and line stoppages as a result of supplier defects.

quality circle A small group of people who normally work as a unit and meet frequently for the purpose of uncovering and solving problems concerning the quality of items produced, process capability or process control. *Syn:* quality control circle; self-directed work group.

quality control The function of verifying conformance to requirements.

quantity discount An allowance determined by the quantity or value of a purchase.

queue A waiting line. In manufacturing, the jobs at a given work center waiting to be processed. As queues increase, so do average queue time and work-in-process inventory.

rang The statistical term referring to the spread in a series of observations. For example, the anticipated demand for a particular product might vary from a low of 10 to a high of 500 per week. The range would, therefore, be 500 – 10 or 490.

rated capacity Capacity calculated from such data as utilization and efficiency, hours planned to be worked, and so on. *Syn:* theoretical capacity. *Ant:* demonstrated capacity.

raw in process (RIP) Inventory kept on the shop floor but not issued to any work order or product. Free stock. Usually a segregating of stockroom inventory from floor inventory accomplished by transferring location from stockroom to shop location. Backflushing or postdeduct relieves the RIP location.

raw material Purchased items or extracted materials that will be converted by the manufacturing process into components and/or products.

receiving This function includes the physical receipt of material; the inspection of the shipment for conformance with the purchase order (quantity and damage); identification and delivery to destination; and preparing receiving reports.

record accuracy The conformity of recorded values in a bookkeeping system to the actual values. For example, the onhand balance of an item maintained in a computer record relative to the actual onhand balance of the items in the stockroom.

regeneration MRP An MRP processing approach where the master production schedule is totally reexploded down through all bills of material, to maintain valid priorities. New requirements and planned orders are completely "regenerated" at that time. *Ant:* net change MRP; requirements alteration.

released order *See* open order.

reorder point *See* order point.

reorder quantity (1) In an order point system of inventory planning, the fixed quantity which should be ordered each time the available stock (onhand plus onorder) falls below the order point. (2) In a variable reorder quantity system, the amount ordered from time period to time period will vary. *Syn:* replenishment order quantity. *See* economic order quantity; lot size.

repetitive manufacturing Production of discrete units, planned and executed to a schedule, usually at relatively high speeds and volumes. Material tends to move in a continuous flow during production, but different items may be produced sequentially within that flow. *See* Just-In-Time.

replenishment lead time The total time that elapses from the moment it is determined that a product is to be reordered until the product is back on the shelf available for use.

replenishment period The time between successive replenishment orders (i.e., replenishment interval). *See* replenishment lead time.

requirements explosion The process of calculating the demand for the components of a parent item by multiplying the parent item requirements by the component usuage quantity specified in the bill of material. *See* dependent demand; gross requirements; material requirements planning.

resource Anything required for production of product whose lack of availability would cause failure to meet the plan.

return on investment A financial measure of the relative return from an investment, usually expressed as a percentage or earnings produced by an asset to the amount invested in the asset.

review period The time between successive evaluations of inventory status to determine whether to reorder. *See* lead time.

rough cut capacity planning The process of converting the production plan and/or the master production schedule into capacity needs for key resources; manpower, machinery, warehouse space, suppliers' capabilities, and, in some cases, money. Product load profiles are often used to accomplish this. *Syn:* resource requirements planning. *See* capacity requirements planning.

routing A set of information detailing the method of manufacture of a particular item. It includes the operations to be performed, their sequence, the various work centers to be involved, and the standards for setup and run. In some companies, the routing also includes information on tooling, operator skill levels, inspection operations, testing requirements, and the like.

run time The planned standard time to produce one or multiple units of an item in an operation. The actual time taken to produce one piece may vary from the standard, but the latter is used for loading purposes and is adjusted to actual by dividing by the appropriate work center efficiency factor.

safety lead time An element of time added to normal lead time for the purpose of completing an order in advance of its real need date. When used, the MRP system, in offsetting for lead time, will plan both order release and order completion for earlier dates than it would otherwise. *See* safety stock.

safety stock (1) In general, a quantity of stock planned to be in inventory to protect against fluctuations in demand and/or supply. (2) In the context of master production scheduling, safety stock can refer to additional inventory and/or capacity planned as protection against forecast errors and/or short-term changes in the backlog. Sometimes referred to as "overplanning" or a "market hedge." *Syn:* buffer stock. *See* hedge.

scheduled receipt Within MRP, open production orders and open purchase orders are considered as "scheduled receipts." On their due date, they will be added to the projected available balance during the netting process for the time period in question. Scheduled receipt dates and/or quantities are not normally altered automatically by the MRP system. Further, scheduled receipts are not exploded into requirements for components, because MRP logic assumes that all components for the manufacture of the item in question have been either allocated or issued to the shop floor. *See* planned order; firm planned order.

scheduler A general term that can refer to material planner, dispatcher, or a combined function.

scheduling The act of creating a schedule, such as a master production schedule, shop schedule, maintenance schedule, vendor schedule.

scheduling rules Basic rules that can be used consistently in a scheduling system. Scheduling rules usually specify the amount of calendar time to allow for a move and for queue, how load will be calculated, and so on.

scrap Material outside of specifications and of such characteristics that rework is impractical.

scrap factor A percentage factor in the product structure used to increase gross requirements to account for anticipated loss within the manufacture of a particular product. *See* shrinkage factor.

service parts Parts used for the repair and/or maintenance of an assembled product. Typically they are ordered and shipped at a date later than the shipment of the product itself.

service parts demand The need for a component to be sold by itself, as opposed to being used in production to make a higher-level product. *Syn:* repair parts demand; spare parts.

setup The process of changing dies or other parts of a machine to product a new part or product. *Syn:* changeover.

setup cost The costs associated with a setup. *Syn:* changeover cost.

setup time The time required for a specific machine, line, or work center to convert from the production of one specific item to another. Typically from the last good part until the next (new) good part. *See* external setup time; internal setup time; major setup; minor setup.

shelf life The amount of time an item may be held in inventory before it becomes unusable.

shipping Provides facilities for the outgoing shipment of parts, products, and components. It includes packaging, marking, weighing, and loading for shipment.

shop calendar A special type of calendar used to facilitate scheduling. It is usually expressed in consecutively numbered working days and excludes weekends, holidays, plant shutdowns. *See* M-day calendar.

shop floor control A system for utilizing data from the shop floor to maintain and communicate status information on shop orders (manufacturing orders) and work centers. The major subfunctions of shop floor control are: (1) assigning priority of each shop order; (2) maintaining work-in-process quantity information; (3) conveying shop order status information to the office; (4) providing actual output data for capacity control purposes; (5) providing quantity by location by shop order for work-in-process inventory and accounting purposes; and (6) providing measurement of efficiency, utilization, and productivity of manpower and machines. *Syn:* production activity control. *See* closed-loop MRP.

shrinkage Reductions of actual quantities of items in stock, in process, in transit. The loss may be caused by scrap, theft, deterioration, evaporation, and so on.

shrinkage factor A percentage factor in the item master record that compensates for expected loss during the manufacturing cycle either by increasing the gross requirements or by reducing the expected completion quantity of planned and open orders. The shrinkage factor differs from the scrap factor in that the former affects all uses of the part and its components, the scrap factor relates to only one usage. *See* scrap factor.

simulation (1) The technique of utilizing representative or artificial data to reproduce in a model various conditions that are likely to occur in the actual performance of a system. Frequently used to test the behavior of a system under different operating policies. (2) Within MRP II, utilizing the operational data to perform "what-if" evaluations of alternative plans, to answer the question "can we do it?" If yes, the simulation can then be run in financial mode to help answer the question "do we really want to?"

single-level backflush A form of backflush which reduces inventory on only the next-level-down parts used in an assembly or subassembly. *See* backflush; superflush.

single-level bill of material A display of those components that are directly used in a parent item. It shows only the relationships one level down.

single minutes exchange of die (SMED) The concept of setup times of less than 10 minutes, developed by Shigeo Shingo in 1970 at Toyota.

SKU Abbreviation for stockkeeping unit.

SPC Abbreviation for statistical process control.

staging Pulling of the material for an order from inventory before the material is required. This action is often taken to identify shortages but can lead to increased problems in availability and inventory accuracy.

standard costs The target costs of an operation, process, or product including labor, material, and overhead charges.

standard deviation A measure of dispersion of data or of a variable. The standard deviation is computed by finding the difference between the average and actual observations, squaring each difference, summing the squared difference, finding the average squared difference (called the "variance") and taking the square root of the variance.

standard error Applied to statistics, such as the mean, to provide a distribution within which samples of the statistics are expected to fall. *See:* standard deviation.

standard house *See* standard time.

standard time The length of time that should be required to (*a*) set up a given machine or operation and (*b*) run one part/assembly/batch/end product through that operation. This time is used in determining machine requirements and labor requirements. Also, frequently used as a basis for incentive payrolls and cost accounting.

start date The date that an order or schedule should be released into the plant based on some form of scheduling rules. The start date should be early enough to allow time to complete the work, but not so early to overload the shop. *See* scheduling rules.

statistical process control (SPC) A quality control methodology that focuses on continuous monitoring during the production process itself, rather than postproduction inspection of the items produced. The intent is to not produce any defective items, by stopping the process before it drifts out of control.

stock (1) Items in inventory. (2) Stored products or service parts ready for sale as distinguished from stores that are usually components or raw materials.

stockkeeping unit (SKU) An item at a particular geographic location. For example, one product stock at six different distribution centers would represent six SKUs, plus perhaps another for the plant at which it was manufactured.

stockless production *See* Just-in-Time.

stockout The lack of materials or components needed. *See* backorder.

stockout costs The lost sale and/or backorder cost incurred as a result of a stockout. *See* stockout.

storage costs A subset of inventory carrying costs, including the cost of warehouse utilities, material handling personnel, equipment maintenance, building maintenance, and security personnel. *See* carrying cost.

subassembly An assembly used at a higher level to make up another assembly. *Syn:* intermediate. *See* component.

suboptimization A problem solution that is best from a narrow points of view but not from a higher or overall company point of view. For example, a department manager who would not work his department overtime to minimize his department's costs may be doing so at the expense of overall company profitability.

summarized bill of material A form of multilevel bill of material, which lists all the parts and their quantities required in a given product structure. Unlike the indented bill of material, it does not list the levels of manufacture and lists a component only once for the total quantity used.

super bill (of material) A type of planning bill, located at the top level in the structure, that ties together various modular bills (and possibly a common parts bill) to define an entire product or product family. The "quantity per" relationship of super bill to modules represents the forecasted percentage of demand of each module. The master scheduled quantities of the super bill explode to create requirements for the modules that also are master scheduled. *See* planning bill; modular bill; common parts bill.

superflush A technique to relieve inventory of all components down to the lowest level using the complete bill of material, based on the count of finished units produced and/or transferred to finished goods inventory. *See* backflush; single-level backflush.

supplier A company or individual that supplies goods or services.

supplier lead time The time that normally elapses between the time an order is received by the supplier and his shipment of the material.

supplier measurement The act of measuring the supplier's performance to the contract. Measurements usually cover delivery, quality, and price.

supplier scheduling A purchasing approach that provides suppliers with schedules, rather than individual hard-copy purchase orders. Normally, a supplier scheduling system will include a business agreement (contract) for each supplier, a weekly (or more frequent) schedule for each supplier extending for some time into the future, and individuals called "supplier schedulers." Also required is a formal priority planning system that works very well, because it is essential in this arrangement to routinely provide the supplier with valid due dates. *See* Just-in-Time; paperless purchasing.

synchronized production A term sometimes used to mean repetitive Just-in-Time production.

target inventory level The equivalent of the "maximum" in a min-max system. The target inventory is equal to the order point plus the order quantity. It is often called an "order up to" inventory level and is used in a periodic review system. May also be the desired or planned inventory level in strategic business planning. *See* min-max system.

teardown time The time taken to remove a setup from a machine or facility. Teardown is an element of manufacturing lead time but often allowed for in setup or run time, rather than separately.

throughput The total volume of production, results, or actions through a facility or department, machine center, work cell, plant, or network of plants.

time bucket The time into which data is summarized. A weekly time bucket would contain all of the relevant data for an entire week. Weekly time buckets are considered to be the most reasonable term to permit effective MRP. *See* bucketless system.

time fence A policy or guideline established to manage where in time various updates or changes to operating plans or schedules take place. For example, changes to the master production schedule can be accomplished easily beyond the cumulative lead time to plan and build any product, whereas changes inside the cumulative lead time becomes increasingly more difficult to a point where changes should be resisted. Time fences can be used to define these points.

time-phased order point (TPOP) An order point technique MRP (or DRP) for independent demand items, where gross requirements come from a forecast, not via explosion generated from a dependent position in a bill of material. It is the basic logic structure of MRP calculations. It starts with onhand inventories, subtracts demand, adds incoming orders, and projects a new onhand balance. This technique can be used to plan distribution center inventories as well as planning for service (repair) parts or supplies. *See* distribution requirements planning (DRP).

time phasing The technique of expressing future sequential demand, supply, and inventories by time period. Time phasing is one of the key elements of material requirements planning.

time series A set of data that is distributed over time, such as demand data in monthly time period occurrences.

time standard The predetermined times allowed for the performance of a specific job. The standard will often consist of two parts; that for machine or operation setup and that for actual work or running time. The standard can be developed through (1) observation of the actual work (time study), (2) summation of standard micromotion times (synthetic time standards, or (3) approximation from the analysis of historical job time.

total productive maintenance Preventive maintenance plus continuing efforts to adapt, modify, and refine equipment to increase flexibility, reduce material handling, and promote continuous flows. It is operator-oriented maintenance with the involvement of all qualified employees in all maintenance activities.

total quality management (TQM) An approach encompassing all phases of an organization, from marketing to delivery, that attempts to insure quality in all things, to ensure that no defective data, processes, or parts are produced in any department. The basic elements include: statistical quality control, process control, quality at the source, line stop, group problem solving, employee involvement, customer satisfaction, team building, and continuous improvement.

TPOP Abbreviation for time-phased order point.

TQC Abbreviation for total quality control. *See* total quality management.

transit time A standard allowance that is assumed on any given order for the physical movement of items from one operation to the next.

transportation inventory Inventory that is in transit between locations. *See* pipeline stock.

turnover (1) *See* inventory turnover. (2) In the United Kingdom and certain other countries, it refers to annual sales volume.

two-bin system A type of order point system in which inventory is carried in two bins. A replenishment quantity is ordered when the first bin is empty. When the material is received, the second bin is refilled and the excess is put into the working bin. This term is also used loosely to describe any order point system even when physical "bins" do not exist. Sometimes used in Just-In-Time manufacturing for small quantities. *See* order point system.

U-lines Production lines shaped like the letter "U." The shape allows workers to easily perform several different and interchangeable tasks without much walking. The number of work stations in a U-line are usually determined by line balancing and the process requirements. U-lines promote communication through cross training and close proximity. *See* cellular manufacturing; group technology.

variable cost An operating cost that varies directly with production volume; for example, materials consumed, direct labor, sales commissions. *Ant:* fixed cost.

variance (1) The difference between the expected (budgeted or planned) and the actual. (2) In statistics, the variance is a measure of dispersion of data.

waste A byproduct of a process or task with unique characteristics requiring special management control. Waste production can usually be planned and somewhat controlled. Scrap is typically not planned and may result from the same production run as waste. In total quality management, waste is any activity, materials, or capacity performed in any company activity that does not add value to the product or service.

where-used A listing of every parent item of any given component, and the respective quantity required, from a bill of material file.

WIP *See* work in process.

work center A specific production facility, consisting of one or more people and/or machines, which can be considered as one unit for purposes of capacity requirements planning and detailed scheduling. *See* cost center.

work in process (WIP) Product or inventory in various stages of completion throughout the plant including raw material that has been released for initial processing, up to completely processed material awaiting final inspection and acceptance as finished product. Many accounting systems also include the value of semifinished stock and components in this category. In Just-In-Time systems, work in process is not used because there are no work orders. Inventory is still in the raw material stock listings until it is backflushed and consumed into the finished product. *See* backflush; postdeduct. *Syn:* in-process inventory.

work order (WO) The authorization order to commit resources to produce or fabricate a product or part. Frequently the term *work order* is used to designate orders to the machine shop for tool manufacture or maintenance. *See* manufacturing order.

work station The assigned location where a worker performs his or her job; it could be a machine or a work bench. *See* work center.

zero inventories A philosophy of manufacturing based on planned elimination of all waste and consistent improvement of productivity. It encompasses the successful execution of all manufacturing activities required to produce a final product, from design engineering to delivery and including all states of conversion from raw material onward. The primary elements of zero inventories are to have only the required inventory when needed; to improve quality to zero defects; to reduce lead times by reducing setup times, queue lengths, and lot sizes; and to incrementally revise the operations themselves to accomplish these things at minimum cost. In the broad sense it applies to all forms of manufacturing, job shop, and process, as well as repetitive. *See* Just-in-Time; stockless production.

Bibliography

Advanced Manufacturing Topics

World Class Manufacturing. Shonberger, R. Free Press, New York, 1986.
Guide to World Class Manufacturing. Buker, David W. The Lowell Press, Kansas City, Mo, 1993.
Manufacturing for Competitive Advantage. Gunn, T.G. Ballinger Publishing, Cambridge, Mass., 1987.
The Goal. Goldratt, E.M., and Cox, J. North River Press, Croton-on-Hudson, N.Y., 1984.
The Race. Goldratt, E.M., and Fox, R.E. North River Press, Croton-on-Hudson, N. Y., 1986.
Journal of Constraints Management. Goldratt, E.M. Avraham Y. Goldratt Institute, New Haven, Conn, 1989.

Capacity Planning

Capacity Management Techniques for Manufacturing Companies with MRP Systems. Wemmerlov, U. American Production and Inventory Control Society, Falls Church, Va., 1984.

Computer Integrated Manufacturing

Computer Integrated Manufacturing and Flexible Manufacturing Systems. Seminar proceedings, American Production and Inventory Control Society, April, Falls Church, Va., 1985.
Computer Integrated Management. Harrington, J.J. Krieger Publishing, New York, 1978.

Cost Accounting Systems

Relevance Lost: The Rise and Fall of Management Accounting. Johnson, H.T., and Robert Kaplan. McGraw-Hill, New York, 1988.
Beyond the Bottom Line: Measuring World Class Performance. McNair, Carol J. Mosconi, William; Norris, Thomas F. Dow Jones-

Irwin, Homewood, Ill., 1989.

Meeting the Technology Challange: Cost Accounting in a JIT Environment. McNair, Carol J., et al. National Association of Accountants, Montvale, N.J., 1988.

Distribution Management

The Distribution Handbook. Robeson, James F., ed. Free Press, New York, 1985.

Distribution Resource Planning. Distribution Management's Most Powerful Tool. Martin, A.J., and Landvater, D.V. CBI Publishing, Boston, 1986.

Strategic Logistics Management. Stock, J.R., and Lambert, D.M. Dow Jones-Irwin, Homewood, Ill., 1987.

The Management of Business Logistics. Coyle, J.J.; Barty, E.J.; and Langley, C.J. 4th ed. West Publishing, Minneapolis, 1988.

Engineering Concepts

Manufacturing Planning and Control Systems. Vollman, T.E.; William Berry; Clay Whybark. 3d. Edition, Richard D. Irwin, Homewood, Ill., 1992.

Bills of Materials. Mather, Hal. Dow Jones-Irwin/APICS, Homewood, Ill., 1986.

Bills of Material: Structured for Excellence. Garwood, R.D. Dogwood Publishing, Marietta, Ga., 1988.

Inventory Planning and Control

Production and Inventory Management. Fogerty, D, and T. Hoffman. McGraw-Hill, New York, 1984.

Production and Inventory Control. 2d., Ed., Plossl, George. McGraw-Hill, New York, 1983.

Just-In-Time Systems

Just-In-Time For America. Wantuck, Kenneth A. The Forum, Ltd., Milwaukee, 1989.

Just-In-Time: Surviving by Breaking Tradition. Goddard, W.E. Oliver Wight Limited Publishing, Essex Junction, Vt., 1986.

Just-In-Time: Making It Happen. Sandras. William A. Jr. Oliver Wight Limited Publishing, Essex Junction, Vt., 1989.
Attaining Manufacturing Excellence. Hall, R.W., Business One Irwin, Homewood, Ill., 1987.
Zero Inventories. Hall, R.W. Dow Jones-Irwin, Homewood, Ill., 1983.
Japanese Manufacturing Techniques. Shonberger, R. Free Press, New York, 1982.
The New Manufacturing Challenge: Techniques for Continuous Improvement. Suzaki, Kiyoshi. Free Press, New York, 1987.

Master Planning

Manufacturing Planning and Control Systems. Vollman, T.E.; William Berry; Clay Whybark. 2d ed., Richard D. Irwin, Homewood, Ill., 1988.
Focus Forecasting. Smith, Bernard. CBI Publishing, Boston, 1984.
Master Production Scheduling - Principles and Practice, Berry, W; Thomas Vollman; and D.Clay Whybark. APICS, Falls Church, Va., 1983.

Material Requirements Planning

Material Requirements Planning. Orlicky, Joseph. McGraw-Hill, New York, 1975.
Manufacturing Resources Planning: MRP II. Wight, Oliver. CBI Publishing, Boston, 1984.

Operations Management

Production and Inventory Control Handbook. Greene, James. McGraw-Hill, New York, 1987.
Operations Management. Shonberger, R. Business Publications, 1985.
Production and Operations Management. Adam, E. and R.

Production Supervision

Understanding the Manufacturing Process. Harrington, J.J. Marcel Dekker, New York, 1984.
Manufacturing for Competitive Advantage. Gunn, T.G. Ballinger Publishing, Cambridge, Mass., 1987.

Production Activity Control

Production Activity Control. Melnyk, S.A., and Carter, P.L. Dow Jones-Irwin, Homewood, Ill., 1987.

Purchasing Management

High Performance Purchasing. Schorr, J.E., and Wallace, T.F. CBI Publishing, Boston, 1986.

Corporate Planning and Procurement. Farmer, D.H., and Taylor, B. John Wiley & Sons, New York, 1975.

Quality Management

The Deming Management Method. Walton, Mary. Putnam Publishing, New York, 1986.

Out of the Crisis. Deming, W. Edwards. MIT Press, Cambridge, Mass., 1986.

Total Quality Control. Feigenbaum, A.V. 3d ed., McGraw-Hill, New York, 1983.

Industrial Quality Control. Charbonneau, H.C., and Webster, G.L. Prentice Hall, 1978.

Juran's Quality Control Handbook. Juran, J.M., and Gryna, F.M., eds. McGraw-Hill, New York, 1988.

Quality, Productivity and Competitive Position. Deming, W.E. QA Productivity Systems, New York, 1983.

Quality Is Free: The Art of Making Quality Certain. Crosby, Philip. McGraw-Hill, New York, 1980.

A TQM Approach to Achieving Manufacturing Excellence. Shores, A. Richard. ASQC Quality Press, Milwaukee, 1990.

Measuring Customer Satisfaction. Hayes, Bob E. American Society of Quality Control, Quality Press, Milwaukee, 1992.

Benchmarking: The Search for Industry Best Practices. Camp, Robert C. ASCQ Quality Press, Milwaukee, 1989.

Statistical Process Control

Statistical Quality Control. Grant, E.L., and Leavenworth, R.S. McGraw-Hill, New York, 1988.

Statistical Process Control:A Guide for Implementation. Berger, R.W., and Hart, T.H. American Society of Quality Control, Milwaukee, 1986.

Addendum

The Just-In Time Self Test—Repetitive — Management Group

This questionnaire is a list of questions about *repetitive manufacturing* operating activities, results or business practices within your companies. Answer each question with simple "yes" or "no" responses. *If there is a substantial activity or effort going on in your company for an item then answer "yes." If no real effort or practice exists answer "no."* This is important to remember because some of the questions may not be able to be answered with a distinct "yes" or "no." If you are not aware of any activity by your company on a particular question answer it "no." You must answer every question.

You must act as objective observers, viewing your operations, people, policies, processes and business practices from a distance. Be careful to avoid ego, pride, or bias in you answers, particularly if you are tied in some way to the activities being questioned. Answer the questions by placing an "X" in the "yes" or "no" columns provided next to each question.

Education and People   Yes   No

1. Management supports and participates in employee involvement programs, that demonstrate trust, delegate authority, and allow autonomous decision making.
2. A continuous and formal training program is in place, which includes new employee indoctrination, skills development, cross training, manufacturing principles, and JIT concepts for all employees.
3. Middle management and supervisors have been reduced and reorganized to support JIT.
4. Key managers, support personnel, and operators have been trained in Just-In-Time practices.
5. Cross-training programs have been implemented with skills tracking and evaluations.
6. Compensation and rewards for employees are based on both employee flexibility and team contribution.
7. Department or team problem-solving groups have been empowered to meet regularly and solve quality and flow problems in their departments.
8. All associated functional departments (engineering, purchasing, marketing, and accounting) are part of companywide problem resolution teams.
9. Management responds to employee ideas and feeds back its responses immediately.

Quality Management

10. Management exhibits consistent support for quality procedures.
11. Inspection sequences have mostly been eliminated and quality in part of the individual operator's responsibility.

Page total

The Just-In Time Self Test—Repetitive (con't)   Management Group

	Yes	No

12. Quality control departments have been replaced with process audit and operator training functions.
13. Early warning statistical quality control tools are in place and are used to monitor and control quality at critical points in the process.
14. Quality errors are repaired or prevented at the source where they occur.
15. Fail-safe (poka-yoke) devices are installed at numerous locations where recurring human quality errors typically occur.
16. Work areas are consistently clean, organized, and free of unnecessary materials and equipment.
17. The majority of incoming materials are certified or source inspected at the suppliers.
18. Supplier quality certification and performance rating programs are in place and continuously monitored.

Factory Flow

19. Transportation networks consistently deliver mixed loads from local and long distance sources.
20. Standardized containers with exact quantities are used between supplier and plant for over 50% of the volume parts.
21. Weekly or daily delivery of 80% or more of production materials are made to the plant and directly to the production line points of use.
22. Supplier deliveries are scheduled by the production processes demand for parts.
23. Standardized containers hold exact consistent quantities and are maintained throughout the shop.
24. Production build and pull quantities are calculated to support only average daily demand.
25. Shop schedules, dispatching, work orders, and expediting have been eliminated and priorities are defined by shift schedules and quantities.
26. Parts are only produced as required by demand and are built in quantities approaching one.
27. Labor is not "kept busy" by building product when not needed at the next operation.

Production Processes

28. Production processes are grouped into product family (group technology) cells or lines.
29. Processes support flexibly mixed model runs with minimum material handling.

Page total ☐ ☐

The Just-In Time Self Test—Repetitive (con't) — Management Group

	Yes	No

30. Manufacturing is actively involved in product and process design improvement for quality and producability.
31. Production parts have been designed to facilitate fast changeover.
32. Tooling and fixtures are available when setups and jobs begin.
33. Rapid setups are established (less than 10 minutes) for most machines and lines.
34. Cycle times of each workstation, cell, or line are matched to upstream and downstream times.
35. Manufacturing processes have been reoriented to eliminate material handling.
36. Process problems are identified and visible signaled immediately on discovery.
37. Manufacturing engineering is located in the production area and is immediately available for problem resolution.
38. Scheduled preventive maintenance is considered an important part of production performance.

Master Planning

39. Daily rate and level schedules are used and meet due dates.
40. MRP is used for demand planning, customer committing, suppliers schedules, and supplier capacity planning.
41. Management participates in the planning and replanning process and commits to a realistic capacity level.
42. Marketing promotes the demonstrated benefits of Just-In-Time to customers.
43. Production rates exactly equal demand rates, or production quantities equal demand quantities.
44. Customer on time delivery rate is 98-plus % as committed.

Purchasing

45. Key volume suppliers are local.
46. Single source suppliers make up greater than 50% of all suppliers.
47. The number of active suppliers has been reduced substantially—by 50% or more.
48. Buyers and suppliers are rated by supplier quality, delivery, and ongoing improvements.
49. Frequent multidepartment contacts are made between suppliers and your plant.

Page total ☐ ☐

The Just-In Time Self Test—Repetitive (con't)   Management Group

                                                                                                                        Yes   No

50. Most paperwork, material handling, transportation, and quality waste has been eliminated between suppliers and plant.
51. Delivery lead time for most parts range from one day to one week.

Data Integrity

52. Inventory record accuracy is 98-plus % or better for both stockrooms and point of use storage.
53. Bills of material accuracy is 99-plus % or better for costing needs and postdeduct inventory.
54. Bills of material flattened—structured with two or less levels in production.
55. Routing detail, methods, or assembly instructions are accurately defined and maintained by timely process flow changes.
56. Clearly defined engineering standards for costing.
57. Shipment forecast variation is ±10% or less by product family in the current time period.
58. Accounting systems and controls have been redesigned to work in a JIT environment.

Results

59. Operational measurements with both targets and tolerances are in place and reviewed daily by management and employees.
60. Overall manufacturing process cycle time and throughput lead times are reduced.
61. Production floor space has been substantially reduced.
62. Work-in-process inventories are continuously reducing.
63. Substantial increases in productivity or shipments per employee.
64. Substantially reduced operating expense.
65. Overall cost of quality is continually reducing.
66. Stockroom inventories are continually reducing.
67. Substantially increased inventories turns.
68. Ontime customer deliveries is continuously increasing for both line fill and order fill rate.

Page total ☐ ☐

Test total ☐ ☐

JIT Self Test Worksheet—Repetitive

Mgm't Level	Scoring Matrix					Level Matrix		
	1–19	20–31	32–44	45–56	57–68	4 Levels	3 Levels	2 Levels
1 Top	1 ___	2 ___	3 ___	4 ___	5 ___	☐	XX	XX
2 Mid	1 ___	2 ___	3 ___	4 ___	5 ___	☐	☐	XX
3 Supv	1 ___	2 ___	3 ___	4 ___	5 ___	☐	☐	☐
1 Line	1 ___	2 ___	3 ___	4 ___	5 ___	☐	☐	☐
					Totals	☐	☐	☐

1. Total all tests from the employees. Sort the tests by management level.
2. Enter the scores from each test by management level in the scoring matrix (use a "hen scratch" like////). For example, if the Supervisory employee's scores for seven employees were 35, 41, 29, 34, 60, 52, 44, then the numbers entered in the scoring matrix would be 1 in 20–31, 4 in 32–44, 1 in 45–56 and 1 in 57–68. Do the same for all management levels in your company.
3. Circle the box 1 through 5 that represents the highest count (hen scratches) in each organization level and enter 1, 2, 3, 4, or 5 in the Level Matrix boxes to the right under the column representing the number of management levels in your organization. Then fill in the appropriate "totals" box at the bottom of the column.

JIT Self Test Rating—Repetitive

Management Levels:	4 Levels	3 Levels	2 Levels		CLASS
	\multicolumn{3}{c}{Worksheet Score Ranges}		
	16–20 ☐	13–15 ☐	9–10 ☐	→	A
Post the total score from the rating worksheet in the appropriate box	12–15 ☐	10–12 ☐	7–8 ☐	→	B
	8–11 ☐	6–9 ☐	5–6 ☐	→	C
	4–7 ☐	3–5 ☐	2–4 ☐	→	D

1. Identify the column representing the number of management levels in your organization.
2. Post the total from the Rating Worksheet in the box representing the range that contains your score.
3. Your JIT Self Test class rating will be in the right hand column!

Individual Rating Scheme—Repetitive

Use this rating scheme if only one individual is being tested.			
Score			
Class A	57 to 68		Well implemented! Don't quit now!
Class B	45 to 56		Some missing elements! Check which items are answered "no" and begin analysis. Prioritize items by cost-benefit then by shortest time to implement.
Class C	32 to 44		Half way there! There are a number of major procedural problems to be solved. More training may be needed. Do not let perceived benefits from JIT slow your attention to process flows and WIP reductions. Investigate support groups.
Class D	20 to 31		You are making some headway! Considerable change to the flows and planning processes will be difficult to overcome. Management support may be lacking or misunderstood. Training in both applications and concepts are an absolute necessity. Support groups will be having real trouble adjusting.
	Below 19		This could be any company. Many of the basic activities for JIT are also necessary for any well-operating firm. The JIT philosophy will require a refocusing of old planning habits, quality control disciplines, demand management, data accuracy, and manufacturing process changes.

Addendum   215

The Just-In Time Self Test—Job Shop   Management Group

This questionnaire is a list of questions about *repetitive manufacturing* operating activities, results or business practices within your companies. Answer each question with simple "yes" or "no" responses. *If there is a substantial activity or effort going on in your company for an item then answer "yes." If no real effort or practice exists answer "no."* This is important to remember because some of the questions may not be able to be answered with a distinct "yes" or "no." If you are not aware of any activity by your company on a particular question answer it "no." You must answer every question.

You must act as objective observers, viewing your operations, people, policies, processes and business practices from a distance. Be careful to avoid ego, pride, or bias in you answers, particularly if you are tied in some way to the activities being questioned. Answer the questions by placing an "X" in the "yes" or "no" columns provided next to each question.

Education and People   Yes   No

1. Management supports and participates in employee involvement programs that demonstrate trust, delegate authority, and allow autonomous decision making.
2. A continuous and formal training program is in place that includes new employee indoctrination, skills development, cross-training, manufacturing principles, and JIT concepts for all employees.
3. Middle management and supervisors have been reduced and reorganized to support JIT.
4. Key managers, support personnel, and operators have been trained in Just-In-Time practices.
5. Cross-training programs have been implemented with skills tracking and evaluations.
6. Compensation and rewards for employees are based on both employee flexibility and team contribution.
7. Department or team problem-solving groups have been empowered to meet regularly and solve quality and flow problems in their departments.
8. All associated functional departments (engineering, purchasing, marketing, and accounting) are part of companywide problem resolution teams.
9. Management responds to employee ideas and feeds back its responses immediately.

Quality Management

10. Management exhibits consistent support for quality procedures.
11. Inspection sequences have mostly been eliminated and quality in part of the individual operator's responsibility.

                                                    Page total

The Just-In Time Self Test—Job Shop (con't)   Management Group

                                                                                                    Yes   No

12. Quality control departments have been replaced with process audit and operator training functions.
13. Early warning statistical quality control tools are in place and are used to monitor and control quality at critical points in the process.
14. Quality errors are repaired or prevented at the source where they occur.
15. Fail-safe (poka-yoke) devices are installed at numerous locations where recurring human quality errors typically occur.
16. Work areas are consistently clean, organized, and free of unnecessary materials and equipment.
17. The majority of incoming materials are certified or source inspected at the suppliers.
18. Supplier quality certification and performance rating programs are in place and continuously monitored.

Factory Flow

19.

20. Standardized containers with exact quantities are used between supplier and plant for over 50% of the volume parts.
21. Weekly or daily delivery of 80% or more of production materials are made to the plant and directly to the production line points of use.
22.

23. Standardized containers hold exact consistent quantities and are maintained throughout the shop.
24.

25.

26. Parts are only produced as required by demand and are built in quantities approaching one.
27. Labor is not "kept busy" by building product when not needed at the next operation.

Production Processes

28. Production processes are grouped into product family (group technology) cells or lines.
29.

                                                                                             Page total

The Just-In Time Self Test—Job shop (con't) — Management Group

	Yes	No

30. Manufacturing is actively involved in product and process design improvement for quality and producability.
31. Production parts have been designed to facilitate fast changeover.
32. Tooling and fixtures are available when setups and jobs begin.
33. Rapid setups are established (less than 10 minutes) for most machines and lines.
34.

35. Manufacturing processes have been reoriented to eliminate material handling.
36. Process problems are identified and visible signaled immediately on discovery.
37. Manufacturing engineering is located in the production area and is immediately available for problem resolution.
38. Scheduled preventive maintenance is considered an important part of production performance.

Master Planning

39.
40. MRP is used for demand planning, customer committing, suppliers schedules, and supplier capacity planning.
41. Management participates in the planning and replanning process and commits to a realistic capacity level.
42. Marketing promotes the demonstrated benefits of Just-In-Time to customers.
43. Production rates exactly equal demand rates, or production quantities equal demand quantities.
44. Customer on time delivery rate is 98-plus % as committed.

Purchasing

45. Key volume suppliers are local.
46. Single source suppliers make up greater than 50% of all suppliers.
47. The number of active suppliers has been reduced substantially—by 50% or more.
48. Buyers and suppliers are rated by supplier quality, delivery, and ongoing improvements.
49. Frequent multidepartment contacts are made between suppliers and your plant.

Page total ☐ ☐

The Just-In Time Self Test—Job Shop (con't)   Management Group

	Yes	No

50. Most paperwork, material handling, transportation, and quality waste has been eliminated between suppliers and your plant.
51.

Data Integrity

52. Inventory record accuracy is 98-plus % or better for both stock rooms and point of use storage.
53. Bills of material accuracy 99-plus % or better for costing needs and postdeduct inventory.
54. Bills of material flattened—structured with two or less levels in production.
55. Routing detail, methods, or assembly instructions are accurately defined and maintained by timely process flow changes.
56. Clearly defined engineering standards for costing.
57. Shipment forecast variation is ±10% or less by product family in the current time period.
58.

Results

59. Operational measurements with both targets and tolerances are in place and reviewed daily by management and employees.
60. Overall manufacturing process cycle time and throughput lead times are reduced.
61. Production floor space has been substantially reduced.
62. Work-in-process inventories are continuously reducing.
63. Substantial increases in productivity or shipments per employee.
64. Substantially reduced operating expense.
65. Overall cost of quality is continually reducing.
66. Stockroom inventories are continually reducing.
67. Substantially increased inventories turns.
68. Ontime customer deliveries is continuously increasing for both line fill and order fill rate.

Page total ☐ ☐

Test total ☐ ☐

JIT Self Test Worksheet—Job Shop

Mgm't Level	Scoring Matrix					Level Matrix		
	1–19	20–31	32–44	45–56	57–68	4 Levels	3 Levels	2 Levels
1 Top	1 ___	2 ___	3 ___	4 ___	5 ___	☐	XX	XX
2 Mid	1 ___	2 ___	3 ___	4 ___	5 ___	☐	☐	XX
3 Supv	1 ___	2 ___	3 ___	4 ___	5 ___	☐	☐	☐
1 Line	1 ___	2 ___	3 ___	4 ___	5 ___	☐	☐	☐
					Totals	☐	☐	☐

1. Total all tests from the employees. Sort the tests by management level.
2. Enter the scores from each test by management level in the scoring matrix (use a "hen scratch" like////). For example, if the Supervisory employee's scores for seven employees were 35, 41, 29, 34, 60, 52, 44, then the numbers entered in the scoring matrix would be 1 in 20–31, 4 in 32–44, 1 in 45–56 and 1 in 57–68. Do the same for all management levels in your company.
3. Circle the box 1 through 5 that represents the highest count (hen scratches) in each organization level and enter 1, 2, 3, 4, or 5 in the Level Matrix boxes to the right under the column representing the number of management levels in your organization. Then fill in the appropriate "totals" box at the bottom of the column.

JIT Self Test Rating—Job Shop

Management Levels:	4 Levels	3 Levels	2 Levels		CLASS
	\multicolumn{3}{c}{Worksheet Score Ranges}		
Post the total score from the rating worksheet in the appropriate box	16–20 □	13–15 □	9–10 □	→	A
	12–15 □	10–12 □	7–8 □	→	B
	8–11 □	6–9 □	5–6 □	→	C
	4–7 □	3–5 □	2–4 □	→	D

1. Identify the column representing the number of management levels in your organization.
2. Post the total from the Rating Worksheet in the box representing the range that contains your score.
3. Your JIT Self Test class rating will be in the right hand column!

Individual Rating Scheme—Job Shop

Use this rating scheme if only one individual is being tested.		
Score		
Class A	57 to 68	Well implemented! Don't quit now!
Class B	45 to 56	Some missing elements! Check which items are answered "no" and begin analysis. Prioritize items by cost-benefit then by shortest time to implement.
Class C	32 to 44	Half way there! There are a number of major procedural problems to be solved. More training may be needed. Do not let perceived benefits from JIT slow your attention to process flows and WIP reductions. Investigate support groups.
Class D	20 to 31	You are making some headway! Considerable change to the flows and planning processes will be difficult to overcome. Management support may be lacking or misunderstood. Training in both applications and concepts are an absolute necessity. Support groups will be having real trouble adjusting.
	Below 19	This could be any company. Many of the basic activities for JIT are also necessary for any well-operating firm. The JIT philosophy will require a refocusing of old planning habits, quality control disciplines, demand management, data accuracy, and manufacturing process changes.

The Just-In Time Self Test—Flow Shop — Management Group

This questionnaire is a list of questions about *repetitive manufacturing* operating activities, results or business practices within your companies. Answer each question with simple "yes" or "no" responses. *If there is a substantial activity or effort going on in your company for an item then answer "yes." If no real effort or practice exists answer "no."* This is important to remember because some of the questions may not be able to be answered with a distinct "yes" or "no." If you are not aware of any activity by your company on a particular question answer it "no." You must answer every question.

You must act as objective observers, viewing your operations, people, policies, processes and business practices from a distance. Be careful to avoid ego, pride, or bias in you answers, particularly if you are tied in some way to the activities being questioned. Answer the questions by placing an "X" in the "yes" or "no" columns provided next to each question.

Education and People   Yes   No

1. Management supports and participates in employee involvement programs, that demonstrate trust, delegate authority, and allow autonomous decision making.
2. A continuous and formal training program is in place, which includes new employee indoctrination, skills development, cross training, manufacturing principles, and JIT concepts for all employees.
3. Middle management and supervisors have been reduced and reorganized to support JIT.
4. Key managers, support personnel, and operators have been trained in Just-In-Time practices.
5. Cross-training programs have been implemented with skills tracking and evaluations.
6. Compensation and rewards for employees are based on both employee flexibility and team contribution.
7. Department or team problem-solving groups have been empowered to meet regularly and solve quality and flow problems in their departments.
8. All associated functional departments (engineering, purchasing, marketing, and accounting) are part of companywide problem resolution teams.
9. Management responds to employee ideas and feeds back its responses immediately.

Quality Management

10. Management exhibits consistent support for quality procedures.
11. Inspection sequences have mostly been eliminated and quality in part of the individual operator's responsibility.

Page total

The Just-In Time Self Test—Flow Shop (con't)   Management Group

                                                                                                 Yes   No

12. Quality control departments have been replaced with process audit and operator training functions.
13. Early warning statistical quality control tools are in place and are used to monitor and control quality at critical points in the process.
14. Quality errors are repaired or prevented at the source where they occur.
15. Fail-safe (poka-yoke) devices are installed at numerous locations where recurring human quality errors typically occur.
16. Work areas are consistently clean, organized, and free of unnecessary materials and equipment.
17. The majority of incoming materials are certified or source inspected at the suppliers.
18. Supplier quality certification and performance rating programs are in place and continuously monitored.

Factory Flow

19. Transportation networks consistently deliver mixed loads from local and long distance sources.
20. Standardized containers with exact quantities are used between supplier and plant for over 50% of the volume parts.
21. Weekly or daily delivery of 80% or more of production materials are made to the plant and directly to the production line points of use.
22. Supplier deliveries are scheduled by the production processes demand for parts.
23. Standardized containers hold exact consistent quantities and are maintained throughout the shop.
24. Production build and pull quantities are calculated to support only average daily demand.
25. Shop schedules, dispatching, work orders, and expediting have been eliminated and priorities are defined by shift schedules and quantities.
26. Parts are only produced as required by demand and are built in quantities approaching one.
27. Labor is not "kept busy" by building product when not needed at the next operation.

Production Processes

28. Production processes are grouped into product family (group technology) cells or lines.
29. Processes support flexibly mixed model runs with minimum material handling.

                                                                  Page total

The Just-In Time Self Test—Flow Shop (con't) — Management Group

	Yes	No

30. Manufacturing is actively involved in product and process design improvement for quality and producability.
31. Production parts have been designed to facilitate fast changeover.
32. Tooling and fixtures are available when setups and jobs begin.
33. Rapid setups are established (less than 10 minutes) for most machines and lines.
34. Cycle times of each workstation, cell, or line are matched to upstream and downstream times.
35. Manufacturing processes have been reoriented to eliminate material handling.
36. Process problems are identified and visible signaled immediately on discovery.
37. Manufacturing engineering is located in the production area and is immediately available for problem resolution.
38. Scheduled preventive maintenance is considered an important part of production performance.

Master Planning

39. Daily rate and level schedules are used and meet due dates.
40. MRP is used for demand planning, customer committing, suppliers schedules, and supplier capacity planning.
41. Management participates in the planning and replanning process and commits to a realistic capacity level.
42. Marketing promotes the demonstrated benefits of Just-In-Time to customers.
43. Production rates exactly equal demand rates, or production quantities equal demand quantities.
44. Customer on time delivery rate is 98-plus% as committed.

Purchasing

45. Key volume suppliers are local.
46. Single source suppliers make up greater than 50% of all suppliers.
47. The number of active suppliers has been reduced substantially—by 50% or more.
48. Buyers and suppliers are rated by supplier quality, delivery, and ongoing improvements.
49. Frequent multidepartment contacts are made between suppliers and your plant.

Page total ☐ ☐

The Just-In Time Self Test—Flow Shop (con't)   Management Group

	Yes	No

50. Most paperwork, material handling, transportation, and quality waste has been eliminated between suppliers and plant.
51. Delivery lead time for most parts range from one day to one week.

Data Integrity

52. Inventory record accuracy is 98-plus % or better for both stock rooms and point of use storage.
53. Bills of material accuracy is 99-plus % or better for costing needs and postdeduct inventory.
54. Bills of material flattened—structured with two or less levels in production.
55. Routing detail, methods, or assembly instructions are accurately defined and maintained by timely process flow changes.
56. Clearly defined engineering standards for costing.
57. Shipment forecast variation is ±10% or less by product family in the current time period.
58. Accounting systems and controls have been redesigned to work in a JIT environment.

Results

59. Operational measurements with both targets and tolerances are in place and reviewed daily by management and employees.
60. Overall manufacturing process cycle time and throughput lead times are reduced.
61. Production floor space has been substantially reduced.
62. Work-in-process inventories are continuously reducing.
63. Substantial increases in productivity or shipments per employee.
64. Substantially reduced operating expense.
65. Overall cost of quality is continually reducing.
66. Stockroom inventories are continually reducing.
67. Substantially increased inventories turns.
68. Ontime customer deliveries is continuously increasing for both line fill and order fill rate.

Page total ☐ ☐

Test total ☐ ☐

JIT Self Test Worksheet—Flow Shop

Mgm't Level	Scoring Matrix					Level Matrix		
	1–19	20–31	32–44	45–56	57–68	4 Levels	3 Levels	2 Levels
1 Top	1 ___	2 ___	3 ___	4 ___	5 ___	☐	XX	XX
2 Mid	1 ___	2 ___	3 ___	4 ___	5 ___	☐	☐	XX
3 Supv	1 ___	2 ___	3 ___	4 ___	5 ___	☐	☐	☐
1 Line	1 ___	2 ___	3 ___	4 ___	5 ___	☐	☐	☐
					Totals	☐	☐	☐

1. Total all tests from the employees. Sort the tests by management level.
2. Enter the scores from each test by management level in the scoring matrix (use a "hen scratch" like ////). For example, if the Supervisory employee's scores for seven employees were 35, 41, 29, 34, 60, 52, 44, then the numbers entered in the scoring matrix would be 1 in 20–31, 4 in 32–44, 1 in 45–56 and 1 in 57–68. Do the same for all management levels in your company.
3. Circle the box 1 through 5 that represents the highest count (hen scratches) in each organization level and enter 1, 2, 3, 4, or 5 in the Level Matrix boxes to the right under the column representing the number of management levels in your organization. Then fill in the appropriate "totals" box at the bottom of the column.

JIT Self Test Rating—Flow Shop

Management Levels:	4 Levels	3 Levels	2 Levels	CLASS
	Worksheet Score Ranges			
	16–20 ☐	13–15 ☐	9–10 ☐	→ A
Post the total score from the rating worksheet in the appropriate box	12–15 ☐	10–12 ☐	7–8 ☐	→ B
	8–11 ☐	6–9 ☐	5–6 ☐	→ C
	4–7 ☐	3–5 ☐	2–4 ☐	→ D

1. Identify the column representing the number of management levels in your organization.
2. Post the total from the Rating Worksheet in the box representing the range that contains your score.
3. Your JIT Self Test class rating will be in the right hand column!

Individual Rating Scheme—Flow Shop

Use this rating scheme if only one individual is being tested.

Score

Class A	57 to 68	Well implemented! Don't quit now!
Class B	45 to 56	Some missing elements! Check which items are answered "no" and begin analysis. Prioritize items by cost-benefit then by shortest time to implement.
Class C	32 to 44	Half way there! There are a number of major procedural problems to be solved. More training may be needed. Do not let perceived benefits from JIT slow your attention to process flows and WIP reductions. Investigate support groups.
Class D	20 to 31	You are making some headway! Considerable change to the flows and planning processes will be difficult to overcome. Management support may be lacking or misunderstood. Training in both applications and concepts are an absolute necessity. Support groups will be having real trouble adjusting.
	Below 19	This could be any company. Many of the basic activities for JIT are also necessary for any well-operating firm. The JIT philosophy will require a refocusing of old planning habits, quality control disciplines, demand management, data accuracy, and manufacturing process changes.

Addendum

The Just-In Time Self Test—Project Management Group

This questionnaire is a list of questions about *repetitive manufacturing* operating activities, results or business practices within your companies. Answer each question with simple "yes" or "no" responses. *If there is a substantial activity or effort going on in your company for an item then answer "yes." If no real effort or practice exists answer "no."* This is important to remember because some of the questions may not be able to be answered with a distinct "yes" or "no." If you are not aware of any activity by your company on a particular question answer it "no." You must answer every question.

You must act as objective observers, viewing your operations, people, policies, processes and business practices from a distance. Be careful to avoid ego, pride, or bias in you answers, particularly if you are tied in some way to the activities being questioned. Answer the questions by placing an "X" in the "yes" or "no" columns provided next to each question.

Education and People   Yes   No

1. Management supports and participates in employee involvement programs, that demonstrate trust, delegate authority, and allow autonomous decision making.
2. A continuous and formal training program is in place, which includes new employee indoctrination, skills development, cross training, manufacturing principles, and JIT concepts for all employees.
3. Middle management and supervisors have been reduced and reorganized to support JIT.
4. Key managers, support personnel, and operators have been trained in Just-In-Time practices.
5. Cross-training programs have been implemented with skills tracking and evaluations.
6. Compensation and rewards for employees are based on both employee flexibility and team contribution.
7. Department or team problem-solving groups have been empowered to meet regularly and solve quality and flow problems in their departments.
8. All associated functional departments (engineering, purchasing, marketing, and accounting) are part of companywide problem resolution teams.
9. Management responds to employee ideas and feeds back its responses immediately.

Quality Management

10. Management exhibits consistent support for quality procedures.
11. Inspection sequences have mostly been eliminated and quality in part of the individual operator's responsibility.

Page total

The Just-In Time Self Test—Project (con't)   Management Group

                                                                                                 Yes   No

12. Quality control departments have been replaced with process audit and operator training functions.
13. Early warning statistical quality control tools are in place and are used to monitor and control quality at critical points in the process.
14. Quality errors are repaired or prevented at the source where they occur.
15. Fail-safe (poka-yoke) devices are installed at numerous locations where recurring human quality errors typically occur.
16. Work areas are consistently clean, organized, and free of unnecessary materials and equipment.
17. The majority of incoming materials are certified or source inspected at the suppliers.
18.

Factory Flow

19.

20.

21.

22. Supplier deliveries are scheduled by the production processes demand for parts.
23.

24.

25.

26. Parts are only produced as required by demand and are built in quantities approaching one.
27. Labor is not "kept busy" by building product when not needed at the next operation.

Production Processes

28. Production processes are grouped into product family (group technology) cells or lines.
29.

                                                                                          Page total

The Just-In Time Self Test—Project (con't)   Management Group

	Yes	No

30. Manufacturing is actively involved in product and process design improvement for quality and producability.
31. Production parts have been designed to facilitate fast changeover.
32. Tooling and fixtures are available when setups and jobs begin.
33. Rapid setups are established (less than 10 minutes) for most machines and lines.
34. Cycle times of each workstation, cell, or line are matched to upstream and downstream times.
35. Manufacturing processes have been reoriented to eliminate material handling.
36. Process problems are identified and visible signaled immediately on discovery.
37. Manufacturing engineering is located in the production area and is immediately available for problem resolution.
38. Scheduled preventive maintenance is considered an important part of production performance.

Master Planning

39. Daily rate and level schedules are used and meet due dates.
40. MRP is used for demand planning, customer committing, suppliers schedules, and supplier capacity planning.
41. Management participates in the planning and replanning process and commits to a realistic capacity level.
42. Marketing promotes the demonstrated benefits of Just-In-Time to customers.
43. Production rates exactly equal demand rates, or production quantities equal demand quantities.
44. Customer on time delivery rate is 98-plus% as committed.

Purchasing

45. Key volume suppliers are local.
46. Single source suppliers make up greater than 50% of all suppliers.
47. The number of active suppliers has been reduced substantially—by 50% or more.
48. Buyers and suppliers are rated by supplier's quality, delivery, and ongoing improvements.
49. Frequent multidepartment contacts are made between suppliers and your plant.

                                                                            Page total

The Just-In Time Self Test—Project (con't)   Management Group

                                                                                                   Yes   No

50. Most paperwork, material handling, transportation, and quality waste has been eliminated between suppliers and plant.
51. Delivery lead time for most parts range from one day to one week.

Data Integrity

52. Inventory record accuracy is 98-plus % or better for both stock rooms and point of view storage.
53. Bills of material accuracy 99-plus % or better for costing needs and postdeduct inventory.
54. Bills of material flattened—structured with two or less levels in production.
55. Routing detail, methods, or assembly instructions are accurately defined and maintained by timely process flow changes.
56. Clearly defined engineering standards for costing.
57. Shipment forecast variation is ±10 % or less by product family in the current time period.
58.

Results

59. Operational measurements with both targets and tolerances are in place and reviewed daily by management and employees.
60. Overall manufacturing process cycle time and throughput lead times are reduced.
61. Production floor space has been substantially reduced.
62. Work-in-process inventories are continuously reducing.
63. Substantial increases in productivity or shipments per employee.
64. Substantially reduced operating expense.
65. Overall cost of quality is continually reducing.
66. Stockroom inventories are continually reducing.
67. Substantially increased inventories turns.
68. Ontime customer deliveries is continuously increasing for both line fill and order fill rate.

Page total

Test total

JIT Self Test Worksheet—Project

Mgm't Level	Scoring Matrix					Level Matrix		
	1–19	20–31	32–44	45–56	57–68	4 Levels	3 Levels	2 Levels
1 Top	1	2	3	4	5	☐	XX	XX
2 Mid	1	2	3	4	5	☐	☐	XX
3 Supv	1	2	3	4	5	☐	☐	☐
1 Line	1	2	3	4	5	☐	☐	☐
					Totals	☐	☐	☐

1. Total all tests from the employees. Sort the tests by management level.
2. Enter the scores from each test by management level in the scoring matrix (use a "hen scratch" like////). For example, if the Supervisory employee's scores for seven employees were 35, 41, 29, 34, 60, 52, 44, then the numbers entered in the scoring matrix would be 1 in 20–31, 4 in 32–44, 1 in 45–56 and 1 in 57–68. Do the same for all management levels in your company.
3. Circle the box 1 through 5 that represents the highest count (hen scratches) in each organization level and enter 1, 2, 3, 4, or 5 in the Level Matrix boxes to the right under the column representing the number of management levels in your organization. Then fill in the appropriate "totals" box at the bottom of the column.

JIT Self Test Rating—Project

Management Levels:	4 Levels	3 Levels	2 Levels		CLASS
	\multicolumn{3}{c}{Worksheet Score Ranges}		
Post the total score from the rating worksheet in the appropriate box	16–20 ☐	13–15 ☐	9–10 ☐	→	A
	12–15 ☐	10–12 ☐	7–8 ☐	→	B
	8–11 ☐	6–9 ☐	5–6 ☐	→	C
	4–7 ☐	3–5 ☐	2–4 ☐	→	D

1. Identify the column representing the number of management levels in your organization.
2. Post the total from the Rating Worksheet in the box representing the range that contains your score.
3. Your JIT Self Test class rating will be in the right hand column!

Individual Rating Scheme—Project

Use this rating scheme if only one individual is being tested.		
Score		
Class A	57 to 68	Well implemented! Don't quit now!
Class B	45 to 56	Some missing elements! Check which items are answered "no" and begin analysis. Prioritize items by cost-benefit then by shortest time to implement.
Class C	32 to 44	Half way there! There are a number of major procedural problems to be solved. More training may be needed. Do not let perceived benefits from JIT slow your attention to process flows and WIP reductions. Investigate support groups.
Class D	20 to 31	You are making some headway! Considerable change to the flows and planning processes will be difficult to overcome. Management support may be lacking or misunderstood. Training in both applications and concepts are an absolute necessity. Support groups will be having real trouble adjusting.
	Below 19	This could be any company. Many of the basic activities for JIT are also necessary for any well-operating firm. The JIT philosophy will require a refocusing of old planning habits, quality control disciplines, demand management, data accuracy, and manufacturing process changes.

Index

A

Accounting systems, design of, 130-32

B

Bills of material (BOM)
  accuracy, 125-26
  chart comparing methods, 127
  phantom BOM, use of, 127-28
  purposes for, 126
  structuring of, 126-28

C

Cause/effect diagrams, quality control through, 48-50
Changeover processes; see Setup processes
Check sheets, quality control through, 46-47
Communication between departments, 34
Compensation and rewards for employees
  performance as factor, 32
  restrictions on, 32
  teamwork, attached to, 32
  trade unions as factor, 32
  training, attached to, 32
Conrad, Jack, 65
Content of test
  data integrity, 16
  education and people; see Education and people
  factory flow, 14-15
  master planning, 15-16
  production processes, 15
  purchasing, 16
  quality management, 13-14
  results of current efforts, 16-17
  staffing; see Education and people
Costs and expenses
  accounting systems, 130-32
  carrying costs defined, 141
  engineering standards for costing, 129
  inventory reduction, impact of, 142
  job costing, 132
  labor costs, relative importance of, 78-80
  materials costs, reduction of, 78-80
  operating expense, reduction of, 141-42
  quality management, of, 41
  setup processes, cost reduction for, 91
  small lots, cost-effective delivery of, 71-72
  supplier quality certification programs, cost-benefit analysis of, 65-66
  training evaluation system, cost effectiveness of, 24
Cross-training
  advantages of, 30
  chart for tracking of, 30-31
  compensation for, 32
  guidelines, 30
  mixed-model production, for, 84
  rewards for, 32
  software for tracking of, 30-31
  tracking of, 30-31
  training evaluation system, as part of, 24
Customer relations
  on-time delivery rate, increases in, 111-12, 145-46
  promotion of JIT as part of, 109-11

Cycle time analysis
  cell cycle times, 92-94
  downstream times, 92-94
  engineering, role of, 104
  flow chart example, 93
  generally, 33
  intercell balancing chart, 94
  line cycle times, 92-94
  manufacturing process cycle time, reduction of, 138
  master planning, as part of, 104-5
  measurement methods, 137
  new product introduction cycle, chart of, 86
  short cycle manufacturing, 2
  upstream times, 92-94
  workstation cycle times, 92-94

D

Data integrity
  accounting systems, design of, 130-31
  bills of material; see Bills of material (BOM)
  engineering standards for costing, 129
  importance of, 124
  inventory record accuracy, 124-25
  job costing, 132
  labor tracking, 133
  point-of-use storage, inventory records for, 124-25
  raw-in-process inventory, 132-33
  shipment forecast variation, 129-30
  shortages in inventory, 125
  stockrooms, inventory records for, 124-25
  summary, 124
  test questions concerning, 16
Decision-making process, workers part in, 25
Delivery from suppliers; see Supplier delivery
Delivery to customers
  accounting for on-time delivery rates, 111-12
  increases in on-time delivery rates, 145-46
  on-time delivery rates, 111-12, 145-46
Demand
  average daily demand, calculation of, 76-77
  production defined by, 80
  rate of demand, production rates as equal to, 110-11
Deming, Dr. W. Edwards, 63
Downsizing of management, 29

E

Education and people
  concepts, list of, 24
  employee development plan; see Employee development plan
  test questions related to, 13
  training evaluation system; see Training evaluation system
Employee development plan
  compensation and rewards for employees, 32
  cross-training; see Cross-training
  needs analysis for training; see Needs analysis for training
  quality management through operator training, 43-44
  training evaluation system; see

Training evaluation system
Employee errors; see Worker errors
Engineering
   cycle time management, planning for, 104
   engineering change order (ECO), routing and, 128-29
   problem solving by, 97
   standards for costing, 129
   supplier quality certification programs for, 63-64
Engineering change order (ECO), 128-29
Errors by workers; see Worker errors
Expenses; see Costs and expenses

F

Factory flow
   average daily demand, calculation of, 76-77
   demand, production defined by, 80
   dispatch lists, elimination of, 78-79
   generally, 71
   idle time of workers, better uses for, 80-81
   labor costs, relative importance of, 78-80
   lot size as factor, 71-73, 77
   materials costs, reduction of, 78-80
   rate of delivery of production materials, 73-74
   scheduling for delivery of production materials, 74-76
   shop schedules, elimination of, 78-79
   small lots, cost-effective delivery of, 71-72
   standardized containers, delivery in, 72-73, 76
   test questions concerning, 14-15
   work orders, elimination of, 78-79
Final assembly schedule (FAS), 104
Fishbone diagrams, quality control through, 48-50
Floor space for production, reduction of, 138-39
Flow charting, 33
Flow factors
   factory flow; see Factory flow
   manufacturing environment, 7-8, 10
   problem-solving for, 33
   test questions covering, 14-15

G

General Motors, 95
Goldratt, Dr. E. M., 77
Group technology, 82-84

I

Implementation plan
   bottom line impact chart, 149
   case study, examples from, 158-59
   education, 153-55
   examples, 158-59
   generally, 4-5, 148
   leadership, 148-49
   management support, 150
   mission statement, 152-53
   model, defining of, 156
   momentum, creation of, 158
   obstacles, types of, 150-52, 157
   pilot selection and preparation, 155
   resistance, types of, 150-52
   team formation for JIT, 152
   team meeting management, 156-57

training, 153-55
Inspection, reduction of, 42-43
Inventory
  cost impact of inventory reduction, 142
  excess inventory, prevention of, 52
  inventory turns, increases in, 143-44
  levels, chart of, 144
  point-of-use storage, inventory records for, 124-25
  raw-in-process inventory, 132-33
  record accuracy, 124-25
  shortages in inventory, 125
  stockroom inventory, reduction of, 124-25, 143
  work-in-process inventories, reduction of, 140
  zero inventories, 2
ISO 9000 guidelines, 63-64

J

Job costing, 132
Job shops
  demand rate equal to production quantities, 110
  production cells for, 101
  production quantities, demand rates equal to, 110
  type of manufacturing environment, as, 7, 9
Just-in-time (JIT) manufacturing
  defined, 2
Just-in-time self test
  answering techniques, 12
  content of test; see Content of test
  generally, 3
  implementation plan; see Implementation plan
  questions; see Content of test
  results evaluation, 4
  scoring; see Scoring

L

Labor
  compensation and rewards for employees, role in, 32
  cost of labor, relative importance of, 78-80
  decision-making process, workers part in, 25
  idle time of workers, better uses for, 80-81
  productivity per employee, increases in, 140-41
  shipments per employee, increases in, 140-41
  tracking of, 133
  worker errors; see Worker errors
Level schedules, 100, 103-5
Load leveling, 103-5, 111
Lot sizing
  cost-effective delivery of small lots, 71-72
  factory flow, effect on, 71-73, 77
  generally, 33
  minimum lot inventory, accessibility of, 84-85
  purchasing for small lot production, 121

M

Maintenance, problem prevention through, 97
Malcolm Baldrige Award, 25, 30
Management

downsizing of, 29
master planning, participation in, 108-9
middle management, reduction of, 29
Manufacturing environment
 generally, 2, 6
 job shop list for, 7, 9
 process flow as factor, 7-8, 10
 project production as factor, 8-10
 repetitive list for, 6-7
 type-model, identification of, 6-10
Marketing promotion of JIT, 109-11
Master planning
 customer on-time delivery rates, 111-12
 cycle time management, 104-5
 daily rate and level schedules, 100-103
 demand rates, production rates as equal to, 110-11
 final assembly schedule (FAS), 104
 generally, 100
 interface chart, 101
 job shops, production cells for, 101
 level schedules, 100, 103-5
 load leveling, 103-5, 111
 management participation in, 108-9
 marketing promotion of JIT, 109-11
 master production schedule (MPS), 100, 102-3
 master schedule integrity, chart for, 102
 material requirements planning (MRP), 105-7
 mixed-model scheduling, 103-5
 planning router, use of, 101
 production plan, chart for, 102
 production rates, demand rates as equal to, 110-11
 test questions concerning, 15-16
Master production schedule (MPS), 100, 102-3
Material requirements planning (MRP), 105-7
Materials
 costs, reduction of, 78-80
 handling of materials, elimination of, 95
 material requirements planning (MRP), 105-7
 quality management of, 62-63
 rate of delivery of, 73-74
 receiving inspection of, 62-63
 scheduling for delivery of, 74-76
 standardized containers, delivery in, 72-73, 76
McDonnell Douglas Computer Systems Company, 125, 130
Middle management, reduction of, 29
Mixed-model production
 cross-training for, 84
 customer on-time delivery rates, 112
 master planning, scheduling for, 103-5
 production processes for, 84-85

N

National Quality Award, 25, 30
Needs analysis for training, 24
 daily review, 28
 design, 27
 development phase, 27
 implementation, 27-28
 indoctrination agenda, 28
 interview cycle, 28
 new employees, training of, 27-28

purpose of, 26
results, 28
small companies, use in, 28
New employees, training of, 27-28
New product introduction cycle, chart of, 86

P

Packaging operation, poka-yoke devices for, 61
Pareto analysis, quality control through, 47-48
Poka-yoke devices
  advantages of, 59
  amateur workers, solutions to errors by, 58
  changeover facilitated by, 88
  contact type devices, 60-61
  design improvements, worker suggestions as to, 59
  detection of defects through, 59
  examples, 60-61
  forgetfulness by workers, solution for, 58
  identification errors by workers, solution to, 58
  inadvertent errors by workers, solutions for, 58
  misunderstandings by workers, solution for, 58
  non-contact type devices, 60-61
  packaging operation, applicability to, 61
  prediction of defects through, 59
  slowness by workers, errors resulting from, 58
  standards lacking, solution for, 58
  surprise errors, solution for, 58
  willful errors by workers, solutions for, 58
Problem-solving teams
  communication improvement between departments, 34
  coordinator, 35
  feedback from higher management, importance of, 36-37
  liaison, 35
  organization of, 35
  purpose of, 33-34
  recorder, 35
  review of, 36
  sponsor, 34-35
  TAC team, 35
  total quality management (TQM) project team, 34-36
Process audits, use of, 43-44
Process control charts, quality control through, 51-54
Process flow charts, quality control through, 45-46
Producibility checklist, 87
Production processes
  cells for manufacturing, 83-84
  changeover, rapidity of, 88
  cycle time analysis; see Cycle time analysis
  factory flow; see Factory flow
  fixtures, availability of, 88-89
  floor space for production, reduction of, 138-39
  generally, 82
  group technology, use of, 82-84
  maintenance, 97
  manufacturing engineering, problem solving by, 97
  material handling, elimination of, 95
  minimum lot inventory, accessibility

of, 84-85
mixed-model production, 84-85
new product introduction cycle, chart of, 86
poka-yoke devices; see Poka-yoke devices
preventive maintenance, 97
problems, immediate identification of, 95-96
process design improvements, 85-87
producibility checklist, 87
product design improvements, 85-87
product families, use of, 82-84
pull system stop signals, 95
scheduled preventive maintenance, 97
setup; see Setup processes
supplier quality certification pro grams for, 63-64
test questions concerning, 15
tooling, availability of, 88-89
Production rates, demand rates as equal to, 110-11
Pull systems
  generally, 33
  measurement of performance, as part of, 137
  stop signals, 95
Purchasing
  buyers, rating of, 117-18
  contracting for capacity, 121-22
  delivery lead time, 120-22
  frequent deliveries as part of, 122
  inventories as factor, 114
  lead time waste reduction, 120-22
  local suppliers as key, 115
  long-range schedules, 121
  material handling reduction, 119-20
  methods for, 114
  paperwork reduction, 119-20
  partnering as part of, 116
  plant/supplier contact, importance of, 118-19
  rating of suppliers and buyers, 117-18
  reduction in number of suppliers, 117
  setup reduction, 121
  single source suppliers, 115-16
  small lot production, 121
  supplier lead times, 114
  supplier quality certification pro grams for, 63-64
  test questions concerning, 16
  traffic management, 114
  transportation reduction, 119-20
  vendor relationships, 114
  waste, elimination of, 119-20

Q

Quality management
  cause/effect diagrams, use of, 48-50
  check sheets, use of, 46-47
  concepts, list of, 38
  cost of quality, 41, 143
  defined, 39-41
  early warning statistical process controls, 44
  fail-safe devises; see Poka-yoke devices
  fishbone diagrams, use of, 48-50
  implementation plan, importance of, 39
  importance of, 38
  inspection, reduction of, 42-43
  operator performance, evaluation of, 43

Index   245

operator training, use of, 43-44
pareto analysis, use of, 47-48
poka-yoke devices; see Poka-yoke devises
problem-solving teams and, 34-36
process audits, use of, 43-44
process control charts, use of, 51-54
process flow charts, use of, 45-46
process variability as factor, 44-45
receiving inspection of materials, 62-63
rework, reduction of, 56-57
run charts, use of, 55-56
scatter diagrams, use of, 50-51
shipments, lateness prevention for, 52
source of errors, identification of, 56
statistical process control (SPC), 44
strategic quality planning, chart of, 40
supplier quality certification program; see Supplier quality certification programs
test questions concerning, 13-14
trends, control charts identifying, 53
worker errors; see Worker errors
workplace organization, 61-62
Questions on test; see Content of test

R

Reporting and measurement
  categories of measurement, 136
  cost impact of inventory reduction, 142
  cycle time measurement, 137
  daily review of, 135-38
  generally, 135
  manufacturing process cycle time, reduction of, 138
  measure board chart, 136
  operating expense, reduction of, 141-42
  operational measurements, 135-38
  production floor space, reduction of, 138-39
  productivity per employee, increases in, 140-41
  quality, reduction in cost of, 143
  shipments per employee, increases in, 140-41
  supplier on-time delivery chart, 139
  target lines for, 137
  throughput, increases in, 140-41
  throughput lead time, reduction of, 138
  tolerance line for, 137-38
  work-in-process inventories, reduction of, 140
Rework, reduction of, 56-57
Routing
  accuracy of, 128-29
  engineering change order (ECO) and, 128-29
  planning router, use of, 101
  streamlining, 128-29
Run charts, quality control through, 55-56

S

Safetran Systems Corporation
  minimum lot inventory, timeliness and accessibility of, 84-85
  supplier quality certification programs, 65
Scatter diagrams, quality control through, 50-51

Scheduled preventive maintenance, 97
Scoring
   company evaluation, 18
   generally, 3
   individual evaluation, 18
   rating narratives, 19-20
   rating scheme, 18
Setup processes
   cost reduction for, 91
   examples, 91-92
   fixtures available at start of, 88-89
   guidelines for, 90
   parts designed for speed in, 88
   purchasing for reduction of, 121
   rapid setups, establishment of, 89-91
   setup-teardown card, use of, 88-89
   tooling available at start of, 88-89
   video cameras as diagnostic tools, 90
Set-up reduction, 33
Shingo, Shigeo, 57, 90
Shipments, lateness prevention for, 52
Short cycle manufacturing, 2
Signaling, 33
Small businesses
   cross-training for, 31
   demand rate equal to production quantities, 110
   needs analysis for training, 28
   problem-solving teams, use of, 36
   production quantities, demand rates equal to, 110
Statistical process control (SPC), 44
Stockless production, 2
Supervisors, reduction of need for, 29
Supplier delivery
   rate of delivery of production materials, 73-74
   scheduling for delivery of production materials, 74-76
   small lots, cost-effective delivery of, 71-72
   standardized containers, delivery in, 72-73, 76
Supplier quality certification programs
   award of certification, 68
   certification contracts, 67-68
   certification representative, role of, 64
   certification teams, 64-66
   cost-benefit analysis, 65-66
   criteria, defining of, 66-67
   engineering, 63-64
   importance of, 62-63
   ISO 9000 guidelines, 63-64
   parts for certification, selection of, 67
   production, 63-64
   purchasing, 63-64
   Safetran Systems Corporation, example of, 65
   scheduling considerations, 68
   summary of process, 68-69
   survey performance, 68

T

Teamwork
   compensation and rewards for employees successful in, 32
   training evaluation system, importance in, 25
Throughput
   lead time for, reduction of, 138
   productivity measured by, 140-41
Trade unions, role of, 32
Traffic management, 114

Training evaluation system
   cost effectiveness as factor, 24
   cross-training, 24
   decision-making process, workers part in, 25
   employee involvement programs, 24
   generally, 24
   importance of, 24
   inter-departmental objectives, 24
   intra-departmental objectives, 24
   needs analysis; see Needs analysis for training
   problem resolution, 24
   purpose of, 26
   teamwork, importance of, 25

W

Waste reduction
   generally, 33
   lead time waste reduction, 120-22
   purchasing, role of, 119-20, 122
Worker errors
   amateurs, by, 58
   fail-safe devises, installation of, 56-59
   forgetfulness, 58
   identification errors, 58
   inadvertent errors, 58
   misunderstandings, 58
   poka-yoke devices; see Poka-yoke devices
   slowness as cause, 58
   source of errors, identification of, 56
   standards lacking, 58
   surprise errors, 58
   willful errors, 58

Z

Zero inventories, 2